Nature of the First Cause

The Discovery of What Triggered the Big Bang

By

Shelton W. Riggs, Jr.

Nature of the First Cause

The Discovery of What Triggered the Big Bang

ISBN-13: 978-1-44862-091-3

ISBN-10: 1-44862-091-0

This book is dedicated to

my three beloved children,

Kerry, Gregory and Susie

and their spouses

Michelle, Lisa and Dean

and to my six beloved

grandchildren,

Kerry & Michelle's

Marlena and Lakota,

Gregory & Lisa's

Mitchell and Kayla,

and Dean & Susie's

Jack and Grant

About The Author

Shelton W. Riggs, Jr. earned undergraduate (University of Texas) and graduate (Vanderbilt) degrees in both Physics and Mathematics.

Professionally, he has consulted as both a hardware and software design engineer to numerous Fortune 500 companies for a wide range of scientific applications. He helped solve several scientific problems for US Army, Air Force and Navy.

Other interests include theoretical physics including quantum mechanics, relativistic mechanics and theoretical mathematics (especially the mystery of prime numbers).

Hobbies include dancing, karaoke, juggling, playing keyboards, writing songs, and writing poetry.

Other Works By Author

The Scientific Theory of God – A bridge Between Faith and Physics provides the reader with basic scientific understanding, interpretation, clarification and answers about concepts and beliefs associated with a Supreme Being. These ideas are developed and based on current theory and the standard model of physics. This new basis has revealed surprising relationships between the scientific definitions of both God and man. A model for the behavior of living matter (bioenergy) has been extended to include the behavior of human beings in terms of perception, decision and action. These concepts combined with the operation of short and long-term memory explain both human consciousness and how the mind controls the body. This model also includes how any desired behavior (provided it does not go against survival) may be achieved. This book offers a scientific creation theory and shows how it is compatible with both the big bang as well as evolutionary theory.

An Alternate Lorentz Invariant Relativistic Wave Equation offers an invariant form which differs from both the Dirac equation as well as the Klein-Gordon equation. Unlike, Schrodinger's non-relativistic wave equation, both the Dirac as well as the Klein-Gordon equation predict wave functions which do not collapse when applied to free systems at rest. On the other hand, Schrodinger's equation predicts wave functions that do collapse when applied to free systems at rest. The alternate relativistic wave equation offered by the author follows Schrodinger's philosophy that if a free system is at rest, then it is a particle with a collapsed wave function.

The Origin of the Planck Mass, Planck Length and Planck Time presents solutions of a system composed of two identical photons which are trapped in each other's gravitational field. The solution applies to any pair of identical particles having zero rest mass. Two solutions were derived. One solution was found by treating two photons as point particles. The quantum mechanical solution came about by treating the two photons as waves. In both solutions, the predicted distance between photons was found to be proportional to the Planck length. The period of the photon's orbit was proportional to the Planck time and the mass energy of each photon is proportional to the Planck mass. The concept of a Planck length, Planck mass and Planck time all emerge from this single model.

Primal Proofs offer several proofs that deal with prime numbers. A proof by contradiction of Goldbach's binary conjecture that every even natural number greater than two (2) can be expressed as the sum of two (2) primes is given. A proof of Goldbach's ternary conjecture that all natural numbers greater than five (5) are the sum of three (3) primes via the binary proof is presented. A proof by construction (utilizing the proof of the binary conjecture) of the twin prime conjecture is offered. A proof of the Riemann hypothesis by deduction is presented. A proof that any prime greater than three (3) is the mean of two other primes is presented. A proof is offered that any even number greater than twelve (12) satisfies Goldbach's binary conjecture in a plurality of ways.

Two entangled formulas that generate all the primes beyond the second prime ($P_2 = 3$) are developed and summarized.

Acknowledgments

I acknowledge God for providing all the resources necessary to answer the question of what caused the big bang.

I acknowledge my country for providing me with all my freedoms especially, freedom of speech.

I acknowledge my parents for providing a secure and nurturing environment that initially made my learning fun.

I acknowledge all my teachers especially my science and mathematics teachers.

I acknowledge every author in the reference section of this book for providing both ideas and data.

Preface

This study was undertaken to investigate the triggering mechanism behind the big bang. This paper presents a scientific theory on how the universe started. During this study, it was realized that even in the simplest case, several events (as uncertainty fluctuations) at the Planck quantum scale would have to take place (with respect to energy, momentum, position and time) before the universe could begin.

The author has made only a few assumptions which have led to a mathematical foundation for the basis of dark (negative) energy, as well as the explanation of the accelerated expansion of our material universe. This accelerated expansion turns out to be associated with a positive gravitational potential energy field. This gravitational potential energy is discovered to possess an equivalent positive mass and pervades all space and time (spacetime). Thus, this gravitational field energy is seen to be the origin of dark matter. Moreover, it may offer more insight into the nature of the Higgs boson and the Higgs field.

This first cause theory provides an explanation for the asymmetry which exists between matter and anti-matter in our world. In this theory, the structure of relationships between energy and spacetime are based on the Heisenberg uncertainty principle as well as the conservation of both energy and momentum.

The derived solutions are correct with respect to both assumptions of the special theory of relativity. Quantitative arguments and expressions based on general relativity are also incorporated. This first cause theory utilizes both the principle of equivalence as well as the cosmological principle. First Cause theory offers an alternate view of cosmological inflation and solves the cosmological flatness problem.

The first cause theory also utilizes an alternate relativistic wave equation. This equation is seen to be a minor correction to the Klein-Gordon relativistic wave equation. A corresponding relativistic free particle wave function that satisfies this wave equation is also presented. The wave functions for the three primal systems which caused the big bang are developed on this basis.

In this theory, new relationships have emerged between old as well as new physical definitions. As an example, one such definition involved a gravitational relativistically corrected Newtonian force between two Planck Masses at a separation distance of one Planck length. The result turns out to be, $F_P = GM_P^2/L_P^2 = c^4/G$ where G, M_P, L_P and c are Newton's gravitational constant Planck mass, Planck length and the speed of light respectively. The quantity c^4/G has therefore been defined to be a Planck Force or F_P.

The fundamental physical laws, basic units, physical constants and basic elementary particles have been included after the glossary.

Please refer to the glossary for the definitions, values and symbols for the various physical quantities within this manuscript.

Table of Contents

Chapter 1
Statement of the Problem

W e will begin the task of converting the question "What caused the big bang?" into an appropriate mathematical statement. First, let us take a look at the uncertainty principle.

The Heisenberg
Uncertainty Principle

The Heisenberg uncertainty principle may be expressed by the following equations.

(U1.0) $\Delta E \Delta t \geq \hbar/2$ or

(U1.1) $(\Delta E)(-\Delta t) \leq -\hbar/2$

where E is the total dynamical energy of the system and t is how long the system had energy E.

Equation (U1.0) states that the standard deviation of any system's energy measurements multiplied by the standard deviation of how long (time measurements) the system had that energy must be greater than or at best (under ideal conditions) equal to a constant. This constant is one half of Planck's constant divided by 2π. Symbolically this is written as $\hbar/2$ where $\hbar = h/2\pi$ and h is Planck's constant. Equation (U1.1) is a consequence of equation (U1.0) except that the standard deviation of time measurements has been negated (multiplied by -1) as well as the right hand side constant. This has the effect of reversing the sense of the inequality. Negative time can be visualized by looking at an analog clock in a mirror.

Since Einstein has demonstrated the equivalence of mass and energy by his famous equation

(U1.2) $E = mc^2$

means that

(U1.2.1) $\Delta E = \Delta mc^2$

by which equations

(U1.0) $\Delta E \Delta t \geq \hbar/2$

(U1.1) $(\Delta E)(-\Delta t) \leq -\hbar/2$

become equations

(U2.0) $\Delta m \Delta t \geq \hbar/(2c^2)$ and

(U2.1) $(\Delta m)(-\Delta t) \leq -\hbar/(2c^2)$

Equation (U2.0) equivalently states that the standard deviation of mass measurements multiplied by the standard deviation of corresponding time measurements must be greater than or at best (under ideal conditions) equal to $(\hbar/2)$ divided by the speed of light squared, c^2.

Equation (U2.1) is a consequence of equation (U2.0) except that the standard deviation of the time measurements is negated and the sense of the inequality is reversed which would produce a negative mass deviation. Negative mass can be visualized by considering it to gravitationally repel or push against positive mass (like the north pole of a magnet pushes on the north pole of another magnet).

Along with equation (U1.2) $E = mc^2$, a related quantity called the rest mass energy E_0 is defined by

(U1.3) $E_0 = m_0c^2$

where m_0 is known as the rest mass (the mass when it is not moving).

The momentum, p of a system is defined as its mass times its velocity by equation

(U1.4) $p = mv$

Another form of the uncertainty principle is given by equation

(U3.0) $\Delta(mv)\Delta r \geq \hbar/2$

Equation (U3.0) states that the standard deviation of a system's momentum measurements multiplied by the standard deviation of its position

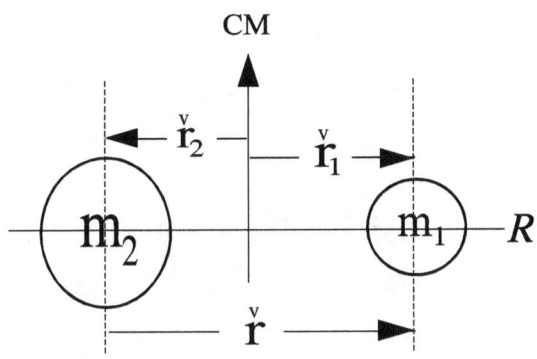

Figure 1 Center of Mass (CM)
Coordinate System
Before the Big Bang

measurements Δr, must be greater than or (even with perfectly ideal measuring devices) equal to $\hbar/2$.

Center of Mass
Coordinate System

Figure 1 shows the center of mass (CM) coordinate system for two masses separated by a vector distance, $\check{r}$. The vector positions of m_1 and m_2 are shown as $\check{r}_1$ and $\check{r}_1$. The horizontal axis R goes through the center of mass of both masses. The vertical axis intersects R at a point that represents the mass of both m_1 and m_2 combined. Figure 1 will be referenced many times throughout the rest of this book.

Positive Newtonian
Gravitational Field Energy

The Newtonian formulation for the motion of two masses $m_1 > 0$, and $m_2 > 0$, acting only under the influence of each other's gravitational field, released from rest (meaning that only radial motion on a line between each of their centers of mass is to be considered) will now be considered.

Referring to figure 1, the equation of motion of m_1 is

(N1.0) $-F = d(m_1 v_1)/dt = -Gm_1 m_2/r^2$

By convention, the force vector on m_1 points left where $-F$ is the force on m_1 due to the gravitational field of m_2, $v_1 < 0$, and m_1 moves left, parallel to the force upon it. The equation of motion of m_2 is

(N1.1) $F = d(m_2 v_2)/dt = Gm_1 m_2/r^2$

The force vector on m_2 points to the right (referring to figure 1) where F is the force on m_2 due to the gravitational field of m_1, $v_2 > 0$ and m_2 moves right, parallel to the force upon it.

The quantity r^2 must be positive. If both masses are positive, the gravitational force, F is attractive and if both masses are negative, the force F, is still attractive. It is only when one mass is positive and the other is negative, that the force F is repulsive, governed by the numerator on the right hand sides of equations (N1.0) and (N1.1). Note, the well known classical gravitational potential energy $V = -Gm_1m_2/r$ between $m_1 > 0$, and $m_2 < 0$, would then be positive.

Below are listed the gravitational force and potential energy for all cases of the signs of m_1 and m_2. Note that the sign of V (the gravitational potential energy) always has the opposite sign as the force. Again, figure 1 defines the coordinate system that will be utilized.

In the case where m_1 and m_2 have opposite signs, the signs of the force and potential energy are governed by the following equations

(N2.0) $F = Gm_1m_2/r^2 < 0$ and

(N2.1) $V = -Gm_1m_2/r > 0$

In the case where m_1 and m_2 have the same signs (are either both positive or both negative), the signs of the force and potential energy are governed by the same following equations, which reverse sign as

(N3.0) $F = Gm_1m_2/r^2 > 0$ and

(N3.1) $V = -Gm_1m_2/r < 0$

Positive Einsteinian Gravitational Field Components

It has been shown by equation (U2.1) $(\Delta m)(-\Delta t) \leq -\hbar/(2c^2)$, that negative matter would be produced during a negative time fluctuation (via the uncertainty principle) before the big bang. Nevertheless, along with appearance of negative matter, an associated positive gravitational field energy is also manifested.

It has been shown that negative gravitational potential energy exists between two positive masses which attract one another. Thus, the existence of positive gravitational potential energy between unlike masses implies that the force is negative and therefore, repulsive.

In Einsteins's "Principle of Relativity" (Dover 0-486-60081-5) Einstein reinforces this conclusion by his expressions for the conservation of energy components of matter and gravitational field. These are given by equation (56) of page 150 of section 17 which is an expression for the conservation of the energy components of matter and gravitational field. Essentially, the result is

(EG1.0) $t^{\sigma}{}_{\mu} + T^{\sigma}{}_{\mu} = $ Constant

where $t^{\sigma}{}_{\mu}$ represent the energy components of the gravitational field and $T^{\sigma}{}_{\mu}$ represents the energy components of positive matter (stress energy momentum tensor). The constant would be representative of the total invariable energy of the system (conservation of energy).

If the total energy is zero, then equation (EG1.0) becomes

(EG2.0) $t^{\sigma}{}_{\mu} + T^{\sigma}{}_{\mu} = 0$ or

(EG2.1) $t^{\sigma}{}_{\mu} = -T^{\sigma}{}_{\mu}$

This means that the energy components of the gravitational field are positive for negative energy or negative matter. This strengthens the idea in the previous section, of equation (N2.1) $V = -Gm_1m_2/r > 0$ where either $m_1 > 0$ and $m_2 < 0$ or $m_1 < 0$ and $m_2 > 0$, showing that the Newtonian gravitational field energy is positive when m_1 and m_2 have opposite signs.

Consequence of Positive Gravitational Field Energy

Consequently, if the signs of the gravitational force equations

(N1.0) $-F = d(m_1v_1)/dt = -Gm_1m_2/r^2$ and

(N1.1) $F = d(m_2v_2)/dt = Gm_1m_2/r^2$

where $m_1 > 0$ and $m_2 > 0$ are reversed by replacing one of the masses with its negative, say $m_2 < 0$, then by definition, the force on m_1 because of m_2 is

(N1.0.R) $F = d(m_1v_1)/dt = -Gm_1m_2/r^2 > 0$

which is positive since $m_2 < 0$, implying $v_1 > 0$. Moreover, the force on m_2 because of m_1 is

(N1.1.R) $-F = d(m_2v_2)/dt = Gm_1m_2/r^2 < 0$

implying $v_2 < 0$, where $m_1 > 0$ and $m_2 < 0$ and therefore repulsive. This in turn, necessarily means that if the force on m_1 due to the gravitational field of m_2 is repulsive, then force on m_2 due to the gravitational field of m_1 is also repulsive by Newton's second law. Note that in this case ($m_1 > 0$ and $m_2 < 0$) by equation (N2.1) $V = -Gm_1m_2/r > 0$, the gravitational potential energy, V is positive.

Relativistic Newtonian Gravitational Force Law

Consider two masses, $m_1 > 0$ and $m_2 > 0$ as in figure 1, which interact because of each other's gravitational field. Conventional usage of these two force equations treat the force as positive for normal gravitational attraction between the two masses. Newton's (relativistically modified) gravitational force on m_1 due to gravitational field of m_2 is again given by equation

(N1.0) $-F = d(m_1v_1)/dt = -Gm_1m_2/r^2$

and similarly the gravitational force on m_2 due to m_1 is

(N1.1) $F = d(m_2v_2)/dt = Gm_1m_2/r^2$

forming the classic action reaction pair. In these equations, both m_1v_1 and m_2v_2 are the respective momenta of both systems. The two masses are separated by a distance r. Note that this force is expressed as the time rate of change in momentum and that all masses are considered a function of their velocities as taught by Einstein's special theory of relativity.

Referring to figure 1, the vector distance, ř between m_1 and m_2 is given by the vector equation

(FC1.0) $\check{r} = \check{r}_1 - \check{r}_2$

where $\check{r}_1$ is the vector position of m_1 and $\check{r}_2$ is the vector position of m_2.

Define scalar equations

(FC1.2) $r_1 = |\check{r}_1|$ and

(FC1.3) $r_2 = |-\check{r}_2|$

where $|\check{r}|$ means the magnitude or length (always positive) of the vector $\check{r}$. This means that

(FC1.1) $r = |\check{r}| = |\check{r}_1 - \check{r}_2| = |\check{r}_1| + |-\check{r}_2| = r_1 + r_2$

The corresponding scalar differential equation is equation

(FC1.4) $dr = |d\check{r}_1| + |-d\check{r}_2| = dr_1 + dr_2$

Recall mass dependence on velocity is given by

(ES1.0) $m = m_0(1- (v/c)^2)^{-1/2}$

and equally, that the velocity dependence on mass is given by

(ES2.0) $v = c(1- (m_0/m)^2)^{1/2}$

In both equation (ES1.0) and (ES2.0) m is the mass at velocity v, m_0 is its rest mass when v is zero. Note that all masses in equations

(N1.0) $-F = d(m_1v_1)/dt = -Gm_1m_2/r^2$ and

(N1.1) $F = d(m_2v_2)/dt = Gm_1m_2/r^2$

must obey equation (ES1.0).

In order to integrate both force equations, a form for the relativistic definition of Newtonian force will be developed. The starting point for this is equation

(ES1.0) $m = m_0(1- (v/c)^2)^{-1/2}$

Taking a time derivative of equation (ES1.0) yields

(ES1.1) $dm/dt = (mv/(c^2 - v^2))dv/dt$

which means linear relativistic acceleration is

(ES1.2) $dv/dt = ((c^2 - v^2)/mv)dm/dt$

and also implys that

(ES1.2.1) $dv/dt \rightarrow 0$ in the limit as $v \rightarrow c$

The definition of relativistic force is

(ES3.0) $F = dmv/dt = mdv/dt + vdm/dt$

and plugging in from equation (ES1.2) results in

(ES3.1) $F = m((c^2 - v^2)/mv)dm/dt + vdm/dt$

and reduces to the relativistic force

(ES3.2) $F = (c^2/v)dm/dt$

where v is assumed to be in the same direction as the force F.

Derivation of mc^2

As a check, note that the classic definition of kinetic energy, T is the integral of force through distance, r written as

(ES3.2.1) $T = \int F dr$

If (ES3.2) is plugged into (ES3.2.1) the result is

(ES3.2.2) $T = \int F dr = \int [(c^2/v) dm/dt] dr$

but since $v = dr/dt$, (ES3.2.2) reduces to

(ES3.2.3) $T = \int F dr = c^2 \int dm$

which upon integration yields

(ES3.2.4) $T = mc^2 - m_0 c^2 = (m - m_0)c^2$

where m is the dynamic mass and m_0 is its rest mass (i.e. when m is not moving). Note that $(-m_0c^2)$ is the integration constant.

If the rest mass energy is interpreted to be potential energy $V = m_0c^2$, the total energy, E_T is defined to be the kinetic energy, T plus the potential energy, V thus,

$$(ES3.2.5) \quad E_T = T + V = mc^2 - m_0c^2 + m_0c^2 = mc^2$$

to formally yield Einstein's famous result. This means that the tool of equation

$$(ES3.2) \quad F = (c^2/v)dm/dt$$

is consistent with special relativity.

Statement Development

Rewriting the force equations

$$(N1.0) \quad -F = d(m_1v_1)/dt = -Gm_1m_2/r^2$$

(N1.1) $F = d(m_2v_2)/dt = Gm_1m_2/r^2$

with the results of equation (ES3.2) yields

(EN1.0) $-F = (c^2/v_1)dm_1/dt = -Gm_1m_2/r^2$ and

(EN1.1) $F = (c^2/v_2)dm_2/dt = Gm_1m_2/r^2$

Note that these last two equations contain the relativistic rates at which two positive masses ($m_1 > 0$ and $m_2 > 0$) gain mass. These mass currents, $dm_1/dt > 0$ and $dm_2/dt > 0$, represent positive mass increases as they each gain velocity under the influence of each other's gravitational force. Both m_1 and m_2 will be treated as functions of their velocities via special relativity.

Both of these last two differential equations represent the relativistic behavior of two positive masses acted upon by their mutual gravitational effects. Recall that their only motion is radial motion (along the line between the masses) since it is assumed that they began their motion starting from rest and separated by some finite non-zero distance.

Thus, these equations represent the statement of the relativistic problem of two positive masses accelerating towards each other because of their mutual gravitation. However, the stated problem doesn't seem to relate to the question of "What caused the big bang?". Could the relativistic gravitational interaction of two positive masses demonstrate anything about the big bang? Even if two positive masses could explain the big bang, one would still have to explain what caused these two masses.

However if equations

(EN1.0) $-F = (c^2/v_1)dm_1/dt = -Gm_1m_2/r^2$ and

(EN1.1) $F = (c^2/v_2)dm_2/dt = Gm_1m_2/r^2$

could be uncoupled and solved, then the solution could be analyzed by changing the sign of one of the masses. As previously discussed, if the sign of one of the masses is changed, then attraction becomes repulsion. Attraction means matter coming together. Repulsion means matter flying apart (i.e. big bang). Thus, there is a chance that these last two equations are related to the creation event after all.

Let us now turn to the preparation and construction of some tools required for their solution.

Chapter 2

Towards a Solution

Note that equations

(FC2.0) $|d\check{r}_1| = dr_1 = -v_1 dt$ since $v_1 < 0$ and

(FC2.1) $|d\check{r}_2| = dr_2 = v_2 dt$

applied to both equations

(EN1.0) $-F = (c^2/v_1)dm_1/dt = -Gm_1m_2/r^2$ and

(EN1.1) $F = (c^2/v_2)dm_2/dt = Gm_1m_2/r^2$

result in equations

(FC2.2) $c^2dm_1 = Gm_1m_2dr_1/r^2$ and

(FC2.3) $c^2dm_2 = Gm_1m_2dr_2/r^2$

Upon adding (FC2.2) and (FC2.3) yields

(FC2.4) $c^2dm_1 + c^2dm_2 = (Gm_1m_2/r^2)(dr_1 + dr_2)$

and by (FC1.4) $dr = |d\check{r}_1| + |{-}d\check{r}_2| = dr_1 + dr_2$
becomes

(FC2.5) $c^2dm_1 + c^2dm_2 = Gm_1m_2dr/r^2$

Solution Integral

Assuming both masses in the center of mass coordinate system have at any instant, a constant ratio, k_R define

(FC3.0) $m_2 = k_Rm_1$ and

(FC3.1) $dm_2 = k_R dm_1$

where k_R is the ratio m_2/m_1. Plugging (FC3.0) and (FC3.1) into (FC2.5) yields

(FC2.6) $(k_R+1)m_1^{-2}dm_1 = (Gk_R/c^2)r^{-2}dr$

whose integral is

(FC2.6.1) $(k_R+1)\int m_1^{-2}dm_1 = (Gk_R/c^2)\int r^{-2}dr$

and upon integration produces

(FC2.7) $(k_R+1)/m_1 = Gk_R/(rc^2) + K_0$

where K_0 is an integration constant to be determined. Plugging back in for k_R from equation (FC3.0) and rearranging yields

(FC4.0) $m_2 m_1^{-2} + m_1^{-1} - Gm_2 m_1^{-1}/(rc^2) = K_0$

Multiplying through by m_1^2 yields

(FC4.1) $m_1 + m_2 - Gm_2m_1/(rc^2) = K_0m_1{}^2$

Solution Interpretation

The left hand side of equation (FC4.1) $\{m_1 + m_2 - Gm_2m_1/(rc^2)\}$ by inspection is the total mass of the system including the $[-Gm_2m_1/(rc^2)]$ term which is the gravitational potential energy, divided by c^2. This is interpreted to be the mass equivalent m_F, of the gravitational potential field energy between m_1 and m_2. Letting

(FC4.2) $m_F = - Gm_2m_1/(rc^2) = V/c^2$

then, equation (FC4.1) becomes

(FC4.3) $m_1 + m_2 + m_F = K_0m_1{}^2$

Multiplying equation (FC4.1) by c^2 yields

(FC4.4) $m_1c^2 + m_2c^2 - Gm_2m_1/r = K_0m_1{}^2c^2$

Note that the left hand side of this equation is

$\{m_1c^2 + m_2c^2 - Gm_2m_1/r\}$ which represents the classic total energy of the system, containing both Einsteinian $[m_1c^2 + m_2c^2]$ as well as Newtonian $[- Gm_2m_1/r]$ components. This fact will later be used to evaluate the integration constant K_0. First, let us examine some pertinent assumptions with respect to the laws of physics before the big bang (creation). These assumptions will help develop more tools needed for the solution to equation

(FC4.4) $m_1c^2 + m_2c^2 - Gm_2m_1/r = K_0m_1^2c^2$

Chapter 3
Before the Big Bang

Quantization of Spacetime

In first cause theory, before the big bang, spacetime is assumed to be quantized. This means that the smallest positive event time is a Planck time T_P. The smallest distance between two points representing any two distinct energy events is a Planck length L_P. Before the big bang, events consist only of positive and negative uncertainty time fluctuations producing both positive matter and negative matter which would, on the average, cancel out. Current quantum mechanics teaches that in a vacuum, positive time fluctuations produce positive matter but only for a   corresponding (via the uncertainty principle) interval of time.

Population Averages Are Zero

The condition of quantized spacetime before the big bang when all material energy of the universe was void and without form will be called vacuum symmetry. Assume that both the total energy E_T, and the total momentum P_T, of the vacuum was on the average equivalent to zero. It is also assumed that the total angular momentum and total charge were both on the average, equivalent to zero.

Single Events and Uncertainty

Assume that before the big bang, all known physical laws were operable, especially conservation of energy as well as the Heisenberg uncertainty principle. The uncertainty principle was previously expressed by equations

(U2.0) $\Delta m \Delta t \geq \hbar/(2c^2)$ and

(U2.1) $(\Delta m)(-\Delta t) \leq -\hbar/(2c^2)$

The standard deviations are denoted in these equations by the Δ symbol. Let us be clear on what the standard deviation means for a single event. The population standard deviation of measurements of mass, m denoted by Δm, is defined by the following equation

(U4.0) $\Delta m = \{[\sum(m_i - m_{av})^2]/N\}^{1/2}$

where m_i is the ith mass measurement and m_{av} is the population average mass of all the events. N is the number of measurements and the sum, $\sum$ goes from $i = 1$ to $i = N$. Since the average energy is assumed to be zero, then so is its equivalent mass average, m_{av}. Only one measurement means only one $m_1 = m$, and $N = 1$. Plugging all this into equation (U4.0) above yields

(U4.1) $\Delta m = m$

which means that the standard deviation of m, Δm is equivalent to m itself.

A similar argument can be made for the time (t) measurement since its population average is also zero. This is because both positive and negative time fluctuations must average to zero since by definition time had not yet began (nothing moved). Thus, we also get

(U4.2) $\Delta t = t$

This means that out of the vacuum before the big bang, a quantum amount of positive or negative mass could materialize, but only for a corresponding quantum amount of positive or negative time. Note also that by equations

(U2.0) $\Delta m \Delta t \geq \hbar/(2c^2)$ and

(U2.1) $(\Delta m)(-\Delta t) \leq -\hbar/(2c^2)$

small amounts of mass can exist for relatively large times (compared to the Planck time, T_P), but large amounts of mass can come into existence only for relatively small times. So long as the amount of mass that materializes is small compared to the Planck mass (M_P), the corresponding amount of gravitational field energy produced will also be small so as not to violate conservation of energy for very long compared to a Planck time, (T_P).

So, before the big bang, whenever a negative time quantum fluctuation produced a negative mass, in order to conserve energy, time would then reverse for another quantum fluctuation to produce positive mass, balance the energy books and reset the Planck clock to zero. This is conservation of energy and the uncertainty principle at work before the big bang. If any energy produced or absorbed was not sustained, then there was no net motion and the flow of time could not begin. Before the big bang, vacuum symmetry (unbroken) ruled, requiring no net spacetime, and net zero mechanical energy.

Three Primal Systems

As implied by both Newtonian and Einsteinian gravitational theory, positive gravitational field energy would exist between negative and positive mass. It is assumed that positive gravitational field energy is associated with a repulsive force.

It follows that initially, before the big bang, any number of positive and negative particles on a quantum scale could be assumed. However, experimentally it is known that a spherically symmetric group of positive masses may be gravitationally represented by their center of mass point. Likewise, it is assumed that a spherically symmetric group of negative masses may be represented by their center of mass point as well.

Therefore, these considerations reduce a many body creation scenario to the initial assumption of only three energy systems namely, a positive mass point, a negative mass point and their associated positive repulsive gravitational field energy in the spacetime between these two masses.

Uncertainty as an Equality

Before the big bang, the Heisenberg uncertainty principle may be expressed as an equality since it will apply to the initial creation event. In this case, the perfect measuring device is the null device (uncreated) and is provided by the ideal conditions of vacuum symmetry. Standard deviations (as has been shown) represent single events and since any additional error caused by the presence of measuring devices are eliminated, the uncertainty principle expressed as an equality become equations

(U2.E.0) $\Delta m \Delta t = \hbar/(2c^2)$ and

(U2.E.1) $(\Delta m)(-\Delta t) = -\hbar/(2c^2)$

We now have enough tools to resume with the analysis of the energy equation

(FC4.4) $m_1 c^2 + m_2 c^2 - G m_2 m_1/r = K_0 m_1^2 c^2$

Chapter 4

Gravitational Symmetry

Evaluation of
Integration Constant

Since it was assumed that the average total mechanical energy E_T before the appearance of m_1 and m_2 was zero, and since $m_1^2 c^2$ is assumed to be non zero, then by induction, $K_0 = 0$ in the total energy equation (FC4.4). Thus, equations (FC4.1), (FC4.3) and (FC4.4) respectively become

(FC5.0) $m_1 + m_2 - Gm_2m_1/(rc^2) = 0$

(FC5.1) $m_1 + m_2 + m_F = 0$ and

(FC5.2) $E_T = m_1c^2 + m_2c^2 - Gm_2m_1/r = 0$

Recall that these last three equations were derived for the case where both masses are positive ($m_1 > 0$ and $m_2 > 0$). The attractive force equations are

(N1.0) $-F = d(m_1v_1)/dt = -Gm_1m_2/r^2 < 0$ and

(N1.1) $F = d(m_2v_2)/dt = Gm_1m_2/r^2 > 0$

Invariance When One Mass Is Negative

If now it is assumed that $m_1 > 0$ and that $m_2 < 0$, then gravitational attraction becomes gravitational repulsion (as discussed in the previous sections on the meaning of positive gravitational field energy) which means that $v_1 > 0$ and $v_2 < 0$. Again referring to figure 1, the repulsive force equations become

(NR1.0) $F = d(m_1v_1)/dt = -Gm_1m_2/r^2 > 0$

(NR1.1) $-F = d(m_2v_2)/dt = Gm_1m_2/r^2 < 0$

keeping in mind that now $m_2 < 0$. Note the dynamic equality of equations (NR1.0) to (N1.0) and (NR1.1) to (N1.1). Now, the gravitational force on m_1 due to m_2 and the motion of m_1 is to the right. Likewise the gravitational force on m_2 due to m_1 and the motion of m_2 is to the left.

Note also the direction of $d\check{r}_1$ and $d\check{r}_2$ as

(NR1.2) $|d\check{r}_1| = dr_1 = v_1dt$ and

(NR1.3) $|d\check{r}_2| = dr_2 = -v_2dt$ since $v_2 < 0$

so applying equation (ES3.2) $F = (c^2/v)dm/dt$, to both force equations

(NR1.0) $F = d(m_1v_1)/dt = -Gm_1m_2/r^2 > 0$ and

(NR1.1) $-F = d(m_2v_2)/dt = Gm_1m_2/r^2 < 0$

yield corresponding equations

(NR2.0) $F = (c^2/v_1)dm_1/dt = -Gm_1m_2/r^2$ and

(NR2.1) $-F = -(c^2/v_2)dm_2/dt = Gm_1m_2/r^2$

since for $v_2 < 0$, $m_2 < 0$, and $m_1m_2 < 0$, must produce $dm_2/dt < 0$. Plugging into these last two equations that

(NR1.2) $|d\check{r}_1| = dr_1 = v_1dt$ and

(NR1.3) $|d\check{r}_2| = dr_2 = -v_2dt$

finally yield equations

(NR1.4) $c^2dm_1 = -Gm_1m_2dr_1/r^2 > 0$ and

(NR1.5) $c^2dm_2 = Gm_1m_2dr_2/r^2 < 0$

which are similar to equations

(FC2.2) $c^2dm_1 = Gm_1m_2dr_1/r^2 > 0$ and

(FC2.3) $c^2dm_2 = Gm_1m_2dr_2/r^2 > 0$

derived under the assumptions that both masses were positive. Subtracting equation (NR1.5) from (NR1.4) results in

(NR1.6) $c^2 dm_1 - c^2 dm_2 = - Gm_1 m_2 dr/r^2$

which when uncoupled and integrated as was equation (FC4.4) $m_1 c^2 + m_2 c^2 - Gm_2 m_1/r = K_0 m_1^2 c^2$ yields,

(NR1.7) $m_1 c^2 - m_2 c^2 + Gm_2 m_1/r = K_1 m_1^2 c^2$

where K_1 is the integration constant. Again, by inspection, the left hand side of (NR1.7) is the total energy, E_T of a system composed of $m_1 > 0$ and $m_2 < 0$. Therefore, as before, setting $K_1 = 0$, equation (NR1.7) finally reduces to

(NR1.8) $E_T = m_1 c^2 - m_2 c^2 + Gm_2 m_1/r = 0$

which looks very similar to equation

(FC5.2) $E_T = m_1 c^2 + m_2 c^2 - Gm_2 m_1/r = 0$

derived for two positive masses. In fact, both equations are the same, since equation (FC5.2) becomes equation (NR1.8) if m_2 is replaced by its negative, $(-m_2)$!

This invariance, can be traced back to the symmetry (much like charge symmetry) that any two masses (regardless of their signs) starting from rest, gravitationally interact to always produce oppositely directed velocities as well as oppositely directed accelerations along a line connecting their centers of mass. Thus, the energy equation

$$(FC5.2) \quad E_T = m_1c^2 + m_2c^2 - Gm_2m_1/r = 0$$

is assumed to be valid at the instant of the big bang. Note that if m_1 is positive and m_2 is negative, then as expected, the gravitational potential energy term, $(-Gm_2m_1/r)$ and its mass equivalent are both positive as verifed by equation (NR1.8). It is now possible to consider the details of the creation event.

Chapter 5

Creation Event

Initial Event Details

Note that by equations

(U2.E.0) $\Delta m \Delta t = \hbar/(2c^2)$ and

(U2.E.1) $(\Delta m)(-\Delta t) = -\hbar/(2c^2)$

a positive vacuum fluctuation whose duration is a Planck time T_P, yields a net half a Planck mass $\frac{1}{2}M_P$ or

(FC6.0) $\Delta m_+ = \hbar/(2c^2 T_P) = \frac{1}{2}M_P$

Similarly, a negative time fluctuation $-T_P$, yields a negative $-\tfrac{1}{2}M_P$ or

(FC6.1) $\Delta m_- = \hbar/(-2c^2 T_P) = -\tfrac{1}{2}M_P$

Recall equations

(FC4.2) $m_F = -Gm_2m_1/(rc^2) = V/c^2$

(FC5.0) $m_1 + m_2 - Gm_2m_1/(rc^2) = 0$

(FC5.1) $m_1 + m_2 + m_F = 0$

(FC5.2) $E_T = m_1c^2 + m_2c^2 - Gm_2m_1/r = 0$

Note that the energy equation (FC5.2) because of (FC4.2) can be rewritten as

(FC6.2) $m_1c^2 + m_2c^2 + m_Fc^2 = 0$

If the energy equation (FC5.2) is solved for m_2, one gets

(FC5.2.1) $m_2 = -m_1/(1 - Gm_1/(rc^2))$

If an initial positive primary mass, $m_{1I} = \frac{1}{2}M_P$ and an initial separation distance from m_{2I} of $r = L_P$ (a Planck length) is substituted into equation (FC5.2.1) and solved for the initial secondary mass, m_{2I}, the result is surprisingly

(FC6.3) $m_{2I} = -M_P$

where m_{1I} and m_{2I} denote the initial amounts of rest mass in the center of mass coordinate system and where the I subscript means initial value. When both m_{1I} and m_{2I} are substituted back into equation (FC5.1) $m_1 + m_2 + m_F = 0$, and solved for the initial field mass, the result is

(FC6.4) $m_{FI} = \frac{1}{2}M_P = m_{1I}$ and

(FC6.5) $m_{2I} + m_{FI} = -m_{1I} = -\frac{1}{2}M_P$

Creation Scenario

All of the above observations makes it possible to formulate the following creation scenario. In Figure 1, assume $m_1 > 0$, $m_F > 0$, $m_2 < 0$ and $|m_2| > |m_F|$. A negative vacuum fluctuation $-T_P$ initially produced m_{2I} and its positive field energy equivalent, m_{FI} simultaneously. So

(FC7.0) $(m_{2I} + m_{FI})(-T_P) = \hbar/(2c^2)$ which means,

(FC7.1) $m_{2I} + m_{FI} = -M_P/2$,

Next, a positive Planck clock tick fluctuation $+T_P$ produced an initial m_{1I} at a distance of one Planck length, $L_P = cT_P$ away from m_{2I}, such that

(FC7.2) $m_{1I}T_P = \hbar/(2c^2)$ thus,

(FC7.3) $m_{1I} = M_P/2$

which resets the Planck clock back to $t = 0$. Note by equation (FC6.3) $m_{2I} = -M_P$, at $t = 0$ that

(FC7.4) $m_{1I} + m_{2I} + m_{FI} = 0 = M_P/2 - M_P/2$

(mass conservation) in agreement with equation (FC6.2) $m_1c^2 + m_2c^2 + m_Fc^2 = 0$ (energy conservation).

So, at $-T_P + T_P = t = 0$, the positive m_{1I} energy plus the equivalent positive gravitational potential energy m_{FI} and a negative mass m_{2I} energy are balanced. The positive repulsive gravitational field energy between m_{2I} and m_{1I} means that m_{2I} and m_{1I} are pushed apart and accelerate away from each other. The force of repulsion is equation (NR1.0) $F = d(m_1v_1)/dt = -Gm_1m_2/r^2$ from which the initial force F_I is calculated to be

(FC7.5) $F_I = GM_P^2/(2L_P^2) = c^4/(2G) = (\frac{1}{2})F_P$

or one half of a Planck force ($\sim 6 \times 10^{43}$ newtons).

Note that even though energy is conserved, vacuum symmetry is broken and since there was energy in motion, the flow of time began and could no longer reverse itself or even need to, since now, energy is automatically balanced.

Before continuing towards the complete analytical solution, let us return to the force equations for $m_1 > 0$ and $m_2 < 0$.

(NR1.0) $F = d(m_1 v_1)/dt = -G m_1 m_2/r^2$ and

(NR1.1) $-F = d(m_2 v_2)/dt = G m_1 m_2/r^2$

Momentum Integral

Adding equations (NR1.0) and (NR1.1) yields

(FC8.0) $d(m_1 v_1)/dt + d(m_2 v_2)/dt = 0$

and whose integral is

(FC8.1) $m_1 v_1 + m_2 v_2 = K_2$

the classic conservation of momentum. Note that K_2 cannot be explicitly time dependent. Note also that for $m_2 < 0$ and $v_2 < 0$, the momentum $m_2 v_2$ is positive.

In the center of mass system, the gravitational repulsive force causes (1) an accelerating separation motion along the line of m_1 and m_2 and at the same time causes by equation

(ES1.0) $m = m_0(1 - (v/c)^2)^{-1/2}$

a relativistic increase in the magnitudes of both positive and negative mass. Thus, the positive repulsive field energy (caused by negative mass) must also increase to conserve energy. Obviously, this is a run away situation, with both systems continuing to accelerate away from each other. Energy, by (FC5.2) $E_T = m_1 c^2 + m_2 c^2 - G m_2 m_1 / r = 0$, continues to be balanced because the positive repulsive gravitational field energy plus the positive mass energy of m_1 is equivalent to the magnitude of the negative mass energy of m_2.

Chapter 6

The Solution

Mass and Position Solution

If the initial ratio $-k_I$ of m_{2I} to m_{1I} as defined by

(FC9.0) $m_{2I} = -k_I m_{1I}$

means that

(FC9.1) $k_I = M_P/(M_P/2) = 2$

If this ratio, k_I at the instant of the big bang, can be shown to be consistent with a similar constant, k_A defined after the creation event, it may lead to the general solutions of the energy equation. Thus, it will now be shown that this ratio, k_I is consistent after the creation event. The solution for r in equation

(FC5.2) $E_T = m_1c^2 + m_2c^2 - Gm_2m_1/r = 0$ is

(FC10.0) $r = Gm_1m_2/[(m_1 + m_2)c^2]$

By inspection, note that if m_2 is negative, then in order for r to remain positive, the magnitude of m_2 must be greater than the magnitude of m_1. Note also, that m_1 and m_2 cannot have equal magnitudes.

Solving equation (FC5.2) for (m_2/m_1) we get

(FC10.1) $m_2/m_1 = -1 + Gm_2/(rc^2)$

and in the center of mass system, assuming a linear relationship between m_1 and m_2, define

(FC10.2) $m_2 = -k_A m_1$

which by equation (FC10.1) produces

(FC10.3) $k_A = 1 - Gm_2/(rc^2)$

Thus, to evaluate k_A, if we put in the initial conditions that $m_{2I} = -M_P$ at $r = L_P$ we get

(FC10.4) $k_A = 1 + GM_P/(L_Pc^2) = 1 + 1 = 2$

which is consistent with the initial ratio, $k_I = 2$. Thus, by equation (FC10.2) $m_2 = -k_Am_1$ means that

(FC10.5) $m_2 = -2m_1$

Therefore, a solution to

(FC10.0) $r = Gm_1m_2/[(m_1 + m_2)c^2]$ is

(FC11.0) $r = 2m_1L_P/M_P$

or equivalently

(FC11.1) $m_1 = rM_P/(2L_P)$

or in terms of m_2, yields equation

(FC11.2) $r = -m_2L_P/M_P$ or

(FC11.3) $m_2 = -rM_P/L_P$

and since by

(FC5.1) $m_1 + m_2 + m_F = 0$, then

(FC11.4) $m_F = rM_P/(2L_P) = m_1$

Equations (FC 10.5, FC11.0, FC11.3 and FC 11.4) above represent the relativistic solutions for m_1, m_2 and m_F in terms of the parameter r which is the separation distance between m_1 and m_2.

Velocity Solution

Thus, let us begin investigating the velocity of m_1 and m_2 because of the general equation

(ES2.0) $v = c(1 - (m_0/m)^2)^{1/2}$

Let us now identify the rest masses of both m_1 and m_2. Recall these initial masses are given by equations

(FC6.3) $m_{2I} = -M_P$ and

(FC6.4) $m_{FI} = \frac{1}{2}M_P = m_{1I}$

where it was assumed that these two masses appeared at rest in the CM system. Thus, these two initial masses are the rest masses. Combining equations

(ES2.0) $v = c(1 - (m_0/m)^2)^{1/2}$ and

(FC6.3) $m_{2I} = -M_P$ implying $m_{02} = -M_P$ and

FC11.3) $m_2 = -rM_P/L_P$ yields,

(FC12.0) $(M_P/m_2)^2 = (1 - (v_2/c)^2) = (L_P/r)^2$

so that

(FC12.1) $(v_2/c)^2 = 1 - L_P^2/r^2$

Again, combining equations

(ES2.0) $v = c(1- (m_0/m)^2)^{1/2}$ and

(FC11.1) $m_1 = rM_P/(2L_P)$ recalling that

(FC6.4) $m_{FI} = \frac{1}{2}M_P = m_{1I}$ implying $m_{01} = \frac{1}{2}M_P$,

yields

(FC12.2) $((M_P/2)/m_1)^2 = (1- (v_1/c)^2) = (L_P/r)^2$

and upon rewriting produces,

(FC12.3) $(v_1/c)^2 = 1 - L_P^2/r^2$

Equating equations

(FC12.1) $(v_2/c)^2 = 1 - L_P^2/r^2$ and

(FC12.3) $(v_1/c)^2 = 1 - L_P^2/r^2$ produces,

(FC12.4) $v_1^2 = v_2^2$

whose solution (by the initial assumptions) is

(FC12.5) $v_1 = -v_2$

which means that as $r \to \infty$, $v_1 \to +c$ and $v_2 \to -c$. If the positive root had been chosen, it would mean that both m_1 and m_2's velocity would be in the same direction. On page 32 of Paul Davies' "The New Physics" (see references), Clifford Will discusses how this has been shown to be impossible.

Note that when $r = L_P$, $v_2 = 0$ and $v_1 = 0$, so ($-M_P$) is consistent with m_2's rest mass and ($M_P/2$) is consistent with m_1's rest mass. Furthermore, by equation (FC11.4) $m_F = rM_P/(2L_P) = m_1$ means that the equivalent rest mass of the gravitational field, m_F is also ($M_P/2$). Moreover, v_2 must be < 0 (since m_2 is moving in the negative direction), in the center of mass of both positive and negative energy systems. Similarly, because of equation (FC12.5) $v_1 = -v_2$ indicates that v_1 has the same magnitude as v_2. Let us see how this is true.

Energy and Momentum Solution

Recall the conservation of momentum equation (FC8.1) $m_1v_1 + m_2v_2 = K_2$. If by equation (FC10.5) $m_2 = -2m_1$ and equation (FC12.5) $v_1 = -v_2$ is plugged into equation (FC8.1), one gets

(FC12.6) $K_2 = 3m_1v_1$

and equation (FC8.1) becomes

(FC12.7) $2m_1v_1 - m_2v_2 = 0$

which can be rewritten as

(FC12.7.1) $m_1v_1 + m_1v_1 - m_2v_2 = 0$

which indicates there is an extra momentum, m_1v_1 that makes the total momentum equal to zero (in accord with the initial assumption that total average momentum P_T was zero before the big bang). Thus, equation (FC12.7.1) becomes clear if it assumed the field energy has momentum

(FC12.8) $m_F v_F = m_1 v_1$

especially when by equation (FC11.4) $m_F = rM_P/(2L_P) = m_1$ would also mean that the velocity of the field is the same as m_1's velocity or

(FC12.8.1) $v_F = v_1$

Thus, the conservation of momentum equation

(FC12.7.1) $m_1 v_1 + m_1 v_1 - m_2 v_2 = 0$

 becomes

(FC13.0) $m_1 v_1 + m_F v_F - m_2 v_2 = 0 = P_T$

and together with the energy equation of

(FC6.2) $m_1 c^2 + m_2 c^2 + m_F c^2 = 0$

rewritten as

(FC13.1) $m_1 c^2 + m_F c^2 + m_2 c^2 = 0 = E_T$

form a symmetric momentum and energy pair.

Note that when $r = L_P$, $v_1 = 0$, so by (FC11.1) m_1 = $rM_P/(2L_P)$, which consistently says that $\frac{1}{2}M_P$ is m_1's rest mass as is the equivalent rest mass of the field energy m_F. Note that v_1 must be > 0 and that v_2 must be < 0 in the center of mass coordinate system of figure 1. Notice also, that when $r = 2L_P$, $v_1 = .867c$ and $v_2 = -.867c$. By equation (ES1.0) $m = m_0(1- (v/c)^2)^{-1/2}$ each mass has become twice its rest mass. So m_2 went from $- M_P$ to $-2M_P$. Similarly, m_1 doubled its mass to M_P and the repulsive field energy equivalent mass m_F at this point ($r = 2L_P$) has also doubled to M_P to conserve energy.

Creation Condition Solved

Thus, to answer the original question of "what caused the big bang?" The answer is that at creation time, $t = 0$, after a negative and positive time fluctuation, three energy systems appear at rest in the center of mass coordinate system of figure 1.

Initially, a positive mass, $m_1 = \frac{1}{2}M_P$, negative mass, $m_2 = -M_P$ and field energy equivalent mass, $V/c^2 = \frac{1}{2}M_P$ have manifested themselves out of the vacuum. The masses are initially separated by a distance, $r = L_P$. Quantum mechanically speaking, three joint primal wave functions appear representing the state of m_1's positive energy, the state of m_2's negative energy and the state of m_F's equivalent gravitational field energy.

As the mathematics has shown, this creation condition set the stage for the big bang. This condition produced a huge force of repulsion and caused energy to be in motion. Energy in motion caused the Planck clock to begin running in a single direction. This obviously gave birth to the spacetime continuum as well as filling it with matter, gravitational field energy and negative matter in motion.

Chapter 7

Solution Analysis

Solution as Schwarzschild Radius

The solution equation for the separation distance (FC11.0) $r = 2m_1L_P/M_P$ is the same as

(FC11.5) $r = 2Gm_1/c^2$ or

(FC11.6) $m_1 = rc^2/(2G)$

Equation (FC11.5) may be considered the dynamic (both m_1 & m_2 are accelerating away from each other) expression for the radius of the universe since it is an expression for the distance between the centers of mass of both m_1 (positive matter of universe) and dark energy m_2 (negative matter of the universe). Moreover, it has the same form as the Schwarzschild radius of a black hole which was associated with one of the few exact analytical solutions of Einstein's gravitational laws of general relativity. Thus, equation (FC11.5) $r = 2Gm_1/c^2$ could be called the Schwarzschild radius of the universe and thus, defines the event horizon of the universe.

Comparison Using the Energy Momentum Vector

A slightly different approach and as a check, would be to use the principle of conservation of four vector momentum on this system. Using equations

(ES1.0) $m = m_0(1- (v/c)^2)^{-1/2}$

(ES2.0) $v = c(1- (m_0/m)^2)^{1/2}$

(U1.2) $E = mc^2$

(U1.3) $E_0 = m_0 c^2$ and

(U1.4) $p = mv$

one may derive equation

(EP1.0) $E^2 - p^2 c^2 = E_0^2$

Equation (EP1.0) can also be derived using the fact that when the energy momentum four vector defined as

(EP1.0.1) $\mathrm{p} = (p, iE/c)$ (where $i = (-1)^{1/2}$)

is squared, the result is the same in all coordinate systems including the rest system where $p = 0$ and $E = E_0$.

Thus, writing equation (EP1.0) for both m_1 and m_2, yield equations

(EP1.1) $E_1^2 - p_1^2 c^2 = E_{10}^2$ and

(EP1.2) $E_2^2 - p_2^2 c^2 = E_{20}^2$

Putting in that $E_2 = -2E_1$ since by (FC10.5) $m_2 = -2m_1$ equation (EP1.2) becomes

(EP1.2.1) $4E_1^2 - p_2^2 c^2 = 4E_{10}^2$

So multiplying equation (EP1.1) by the factor 4 and subtracting (EP1.2.1) produces

(EP1.2.2) $p_2^2 = 4p_1^2$

putting in that $p_2 = -2m_1 v_2$ and $p_1 = m_1 v_1$ since by (FC10.5) $m_2 = -2m_1$, equation (EP1.2.2) becomes

(EP1.3) $v_1^2 = v_2^2$

which is exactly the same as equation

(FC12.4) $v_1^2 = v_2^2$

derived in the velocity solution section of the last chapter.

This means that using the conservation of momentum approach is consistent with the results that were produced by the previous conservation of energy approach. Recall that a velocity solution for all three primal velocities v_1, v_2 and v_F were produced.

Center of Mass System

Let us write down the expression for the center of mass of m_1 and m_2 in a coordinate system located to the right of m_2 as in figure 1.

(CM1) $\check{r}_C = (m_1\check{r}_1 - m_2\check{r}_2)/(m_1+m_2)$

since $\check{r}_2 < 0$ and $m_2 < 0$, and $m_2\check{r}_2 > 0$

where the center of mass coordinate system means that the vertical axis labeled (CM) has been chosen to lie on a line between m_1 and m_2 which crosses the horizontal axis R at a point that represent the center of mass of both m_1 and m_2.

By (FC10.5) $m_2 = -2m_1$ and plugging this into (CM1) results in,

(CM2) $\check{r}_C = -(\check{r}_1 + 2\check{r}_2)$

However since by (FC1.0) $\check{r} = \check{r}_1 - \check{r}_2$, results in

(CM2.1) $\check{r}_1 = -\check{r}_C/3 + 2\check{r}/3$

and

(CM2.2) $\check{r}_2 = -\check{r}_C/3 - \check{r}/3$

Taking a time derivative of equation (CM2) and because of (FC12.5) $v_1 = -v_2$ results in

(CM3) $v_C = -(v_1 + 2v_2) = v_F = v_1$

Note that since v_C is equivalent to v_F the same argument can be made for v_C that was made earlier for v_1 in that $v_1 \rightarrow +c$ and $v_2 \rightarrow -c$ very rapidly. Thus, the center of mass velocity given by (CM3) obeys the following conditions

(CM4) $v_C \rightarrow c$ as $v_1 \rightarrow c$ and $v_F \rightarrow c$ and $v_2 \rightarrow -c$

Constancy of the Repulsive Force

If we plug m_1 and m_2 from equations

(FC11.1) $m_1 = rM_P/(2L_P)$ and

(FC11.3) $m_2 = -rM_P/L_P$

into equation (NR1.0) $F = d(m_1v_1)/dt = -Gm_1m_2/r^2$, the repulsive force on m_1 by m_2 is

(FC14.0) $F = Gr^2(Mp/L_P)^2/(2r^2) = c^4/(2G) = \frac{1}{2}F_P$

This means that the universe will continue to expand unabated, since the constant force of expansion does not depend on m_1 or m_2, time, or their separation distance r. Note that this does not imply that m_1 or m_2 could not themselves expand or contract since m_1 and m_2 represent the respective centers of mass of both the positive and negative energy systems.

Time Dependent Solution

The following equations were derived with r as the dependent variable.

(FC11.0) $r = 2L_P m_1 / M_P$

(FC11.1) $m_1 = r M_P / (2L_P)$

(FC11.3) $m_2 = -r M_P / L_P$

(FC11.5) $r = 2G m_1 / c^2$

(FC11.6) $m_1 = r c^2 / (2G)$

(FC12.1) $(v_2/c)^2 = 1 - L_P^2 / r^2$

(FC12.3) $(v_1/c)^2 = 1 - L_P^2 / r^2$

However, using the addition of velocities law of special relativity to get the relative velocity of m_1 with respect to m_2, v_{rel} results in equation

(FC15.0) $v_{rel} = (v_1 - v_2)/(1 - v_1v_2/c^2) = dr/dt$

taking the limit of v_{rel} as $v_1 \rightarrow c$ and $v_2 \rightarrow -c$ yields

(FC15.1) $v_{rel} = (c+c)/(1+1) = c$

therefore, define

(FC15.2) $t = (r - L_P)/v_{rel} = (r - L_P)/c$

which can be rewritten as

(FC15.3) $r = ct + L_P$

since at $t = 0$, $r = L_P$

Thus, the above equations may be rewritten in time dependent form which are

(FCt11.0) $t = 2m_1 T_P/M_P - T_P$

(FCt11.1) $m_1 = tM_P/(2T_P) + M_P/2$

(FCt11.3) $m_2 = -tM_P/T_P - M_P$

(FCt11.5) $t = 2Gm_1/c^3 - T_P$

(FCt11.6) $m_1 = tc^3/(2G)) + M_P/2$

(FCt12.1) $(v_2/c)^2 = 1 - 1/(t/T_P + 1)^2$

(FCt12.3) $(v_1/c)^2 = 1 - 1/(t/T_P + 1)^2$

Constancy of Mass Currents

Taking the derivatives of equations (FCt11.3) and (FCt11.6) produce

(FCt11.3.1) $dm_2/dt = -M_P/T_P = -c^3/G$ and

(FCt11.6.1) $dm_1/dt = \frac{1}{2}(c^3/G) = dm_F/dt$

Light Speed Expansion

Recall the conservation of energy relation (FC5.2) $m_1c^2 + m_2c^2 - Gm_2m_1/r = 0$. Let it be renumbered for this section as

(CE1) $m_1c^2 + m_2c^2 - Gm_1m_2/r = 0$

Taking a derivative with respect to time yields,

(CE2) $(dm_1/dt)(c^2 - Gm_2/r) + (dm_2/dt)(c^2 - Gm_1/r) + (Gm_1m_2/r^2)(dr/dt) = 0$

Plugging into equation (CE2) from the following four equations

(FC11.1) $m_1 = rM_P/(2L_P)$

(FC11.3) $m_2 = -rM_P/L_P$

(FCt11.3.1) $dm_2/dt = -M_P/T_P = -c^3/G$

(FCt11.6.1) $dm_1/dt = \frac{1}{2}c^3/G = dm_F/dt$

yields

(CE3) $\frac{1}{2}c^3/G(c^2 + GM_P/L_P) - c^3/G(c^2 - GM_P/(2L_P))$
$- c^4/(2G)(dr/dt) = 0$

and upon multiplying out reduces to

(CE4) $\frac{1}{2}c^5/G + \frac{1}{2}c^5/G + -c^5/G + \frac{1}{2}c^5/G$
$- c^4/(2G)(dr/dt) = 0$

and collecting terms becomes

(CE5) $\frac{1}{2}c^5/G - c^4/(2G)(dr/dt) = 0$ or

(CE6) $dr/dt = c$

which is consistent with equation

(FC15.3) $r = ct + L_P$

without having to invoke the relativistic addition of
velocities law.

Chapter 8

Quantum Mechanical Analysis

W hat follows is a quantum mechanical analysis of the three energy systems (positive energy, field energy and negative energy which (in the first cause theory) are the basic components of the material universe.

Free Particle Relativistic Plane Wave

Consider an energy system having a four vector position, $\hat{s}$ as

(B1) $\hat{s} = (x, y, z, ict)$

with initial four vector position as

(B2) $\hat{s}_0 = (0, 0, 0, 0)$

and having a four vector momentum

(B3) $\flat = (p_x, p_y, p_z, iE/c)$

where again p stands for momentum, mv or mass times velocity. The initial four vector momentum is

(B4) $\flat_0 = (0, 0, 0, iE_0/c)$

where c is the velocity of light in vacuum and E is its total energy given by $E = mc^2$. The rest mass energy is given by $E_0 = m_0c^2$. Form the vector scalar product of the two difference vectors $(\hat{s} - \hat{s}_0)$ and $(\flat - \flat_0)$ to get

(B5) $(\hat{s} - \hat{s}_0) \bullet (\flat - \flat_0) = xp_x + yp_y + zp_z - (E - E_0)t$

where • stands for vector scalar product. Now choose the system moving down the x axis so that the right hand side of equation (B5) becomes,

(B6) $(\hat{s} - \hat{s}_0) \bullet (\text{þ} - \text{þ}_0) = xp - (E - E_0)t$

where $p_x = p$.

If the x axis is superimposed on the separation distance (r) between positive energy (m_1) and dark energy (m_2), then (B6) becomes

(B6.1) $(\hat{s} - \hat{s}_0) \bullet (\text{þ} - \text{þ}_0) = rp - (E - E_0)t$

where $p_x = p_r = p$

Such a free particle system can be described by a position and time dependent wave function or wave amplitude, $\Psi(r,t)$ as

(B7.0) $\Psi(r,t) = \Psi_0 e^{\{[rp - (E-E_0)t]i/\hbar\}}$

$\qquad\qquad = \Psi_0 \exp\{[rp - (E-E_0)t]i/\hbar\}$

where exp{} means that what is inside the curly brackets is the exponent of the natural base e. $\hbar$ is again Planck's constant divided by 2π, and again i = $(-1)^{1/2}$. $\Psi_0 = \Psi(0, 0)$ is the initial wave amplitude. Treat the positive and dark energy systems as two waves moving away from each other down the axis of separation distance (r). The gravitational field energy will also be regarded as a wave which expands between these two systems with relative velocity given by

(CM3) $v_C = -(v_1 + 2v_2) = v_F = v_1$

and mass equivalent of the field energy given by equations

(FC4.2) $m_F = - Gm_2m_1/(rc^2) = V/c^2$ and

(FC11.4) $m_F = rM_P/(2L_P) = m_1$

Derivation of the
Total Wave Function

We begin by writing the wave function in (B7.0) for positive, negative and gravitational field energy systems, assuming as usual that $m_1 > 0$ and $m_2 < 0$, thus for m_1

(B7.1) $\Psi_1(+r,t) = \Psi_{10}exp\{[rp_1 - (E_1 - E_{10})t]i/\hbar\}$

where m_1 is moving to the right (+r). Ψ_{10} is the initial wave function of system m_1 and E_{10} is the rest mass energy of m_1. For the m_2 system we get

(B7.2) $\Psi_2(-r,t) = \Psi_{20}exp\{[-rp_2 - (E_2 - E_{20})t]\}i/\hbar$

m_2 moving to the left (−r). For the field energy we get

(B7.1.1) $\Psi_F(+r,t) = \Psi_{F0}exp\{[rp_F - (E_F - E_{F0})t]\}i/\hbar$

positive field energy moving to the right (+r). where, $\Psi_{10}(0,0)$, $\Psi_{20}(0,0)$, and $\Psi_{F0}(0,0)$ are the respective initial wave amplitudes. Therefore the joint probability amplitudes that the universe is composed of positive, gravitational and dark energy systems is

(B7.3) $\Psi_{U3}(r,t) = \Psi_{10}\Psi_{20}\Psi_{F0}\exp\{rp_1-(E_1-E_{10})t -rp_2 -(E_2-E_{20})t + rp_F-(E_F- E_{F0})t\}i/\hbar$

Let $\Psi_{U30}(0,0) = \Psi_{10}(0,0)\Psi_{20}(0,0)\Psi_{F0}(0,0)$ be the initial amplitudes and recalling from the solution that $E_{20} = -2E_{10}$, $E_2 = -2E_1$, $E_F = E_1$ and $E_{F0} = E_{10}$ so that B7.3 becomes

(B7.4) $\Psi_{U3}(r,t) = \Psi_{U30}\exp\{(rp_1-rp_2 + rp_F + E_1t- E_Ft)\}i/\hbar$

Substituting in for $p_1 = m_1v_1$, $p_2 = m_2v_2$, $m_2 = -2m_1$, $E_1 = m_1c^2 = E_F$, $p_F = m_1v_1$ where p_F is the gravitational field momentum, equation (B7.4) becomes

(B7.4.1) $\Psi_{U3}(r,t) = \Psi_{U0} \exp\{(rm_1v_1 - 2rm_1v_1 + rm_1v_1 + E_1t - E_1t\}i/\hbar$

which simplifies into,

(B7.4.2) $\Psi_{U3}(r,t) = \Psi_{U0} \exp\{0i/\hbar\} = \Psi_{U30} = $ constant

Another check is to separate the energy and momentum components and write (B7.3) as

(B7.5) $\Psi_{U3}(r,t) = \Psi_{12F}(r,t) = $
$\Psi_{10}\Psi_{20}\Psi_{F0} \exp\{r[(p_1- p_2 + p_F)] - [(E_1 + E_2 + E_F) - (E_{10} + E_{20} + E_{F0})]t\}i/\hbar$

Conservation of momentum requires that the total momentum, $p_1 - p_2 + p_F$ be zero, and conservation of energy requires that both the total energy, $E_1 + E_2 + E_F$ and total initial energy, $E_{10} + E_{20} + E_{F0}$ be zero. Thus, equation (B7.5) again reduces to

(B7.4.2) $\Psi_{U3}(r,t) = \Psi_{U0} \exp\{0i/\hbar\} = \Psi_{U30} = $ constant

This means that the wave function of the universe is a constant, being the same now as it was before the big bang. This is not too surprising since it was assumed that on the average, the total energy and momentum before the big bang averaged to zero. Moreover, it was assumed that the probability amplitude of the universe consisted of (1) the probability amplitude of positive energy mc^2, (2) the probability amplitude of dark energy $-2mc^2$, and (3) the probability amplitude of positive repulsive gravitational energy equivalent mc^2. This also means that all positive energy, all dark energy and their mutual repulsive gravitational energy are entangled, originating from the two seeds of the original positive (positive field energy and positive matter) and negative (negative matter) Planck pair produced from a normal vacuum fluctuation.

Relativistic Free Particle Wave Equation

It can be shown that the wave function (B7.0) satisfies a relativistically invariant wave equation

(WB7) $(-\hbar^2/(m+m_0))(\nabla^2)\Psi(r,t) = (E-E_0)\Psi(r,t)$

which has a slightly different form than either the Dirac wave equation or the Klein-Gordon wave equation. The Klein Gordon equation is reproduced here for reference as;

(WB8) $(\nabla^2 - \partial^2/(\partial(c^2t^2)))\Psi(r,t) = (m_0^2c^2/\hbar^2)\Psi(r,t) = (E_0^2/(\hbar^2c^2))\Psi(r,t)$

A corresponding free particle wave function that will satisfy the Klein-Gordon wave equation is,

(B7.9) $\Psi(r,t) = \Psi_0\exp(rp - Et)i/\hbar$

It is found that if the above Klein-Gordon wave function (B7.9) is used instead of

(B7.0) $\Psi(r,t) = \Psi_0\exp\{(rp - (E-E_0)t)\}i/\hbar$

for the wave function, the same results of equation

(B7.4.2) $\Psi_{U3}(r,t) = \Psi_{U0} \exp\{0\}i/\hbar = \Psi_{U30} = $ constant

will again emerge. The reason that equation

(B7.0) $\Psi(r,t) = \Psi_0\exp\{(rp - (E-E_0)t)\}i/\hbar$

is preferred over

(B7.9) $\Psi(r,t) = \Psi_0\exp\{(rp - Et)\}i/\hbar$

and that equation

(WB7) $(-\hbar^2/(m+m_0))(\nabla^2)\Psi(r,t) \ = \ (E-E_0)\Psi(r,t)$

is preferred over equation

(WB8) $(\nabla^2 - \partial^2/(\partial(c^2t^2)))\Psi(r,t) = (E_0^2/(\hbar^2c^2))\Psi(r,t)$

is that in (B7.9) when p = 0 (when the particle is at rest), then $E_0 = m_0c^2$ which by

(B7.9) $\Psi(r,t) = \Psi_0\exp\{(rp - Et)\}i/\hbar$,

reduces the Klein – Gordon wave function to

(B7.10) $\Psi(0, t) = \Psi_0\exp\{-E_0t\}i/\hbar$

since $p = 0$. This means that the wave function for any system at rest (except for photons with $E_0 = 0$) does not collapse and will oscillate with respect to time! Note that a corresponding Schrodinger wave function for a free particle contains no such paradox and is represented by equation

(B7.11) $\Psi(r,t) = \Psi_0\exp\{(rp - E_kt)\}i/\hbar$

where $p = m_0v$ and $E_k = p^2/(2m_0) = (½)m_0v^2$

Note that when a Schrodinger system is at rest, then $v = 0$, $p = 0$, the wave function collapses and equation (B7.11) consistently becomes

(B7.11.1) $\Psi(0,0) = \Psi_0$

which says that if nothing moves, nothing waves.

(B7.0) $\Psi(r,t) = \Psi_0 \exp\{(rp - (E-E_0)t)\}i/\hbar$

may be considered as the "corrected" Klein – Gordon wave function of

(B7.9) $\Psi(r,t) = \Psi_0 \exp\{(rp - Et)\}i/\hbar$

Similarly, equation

(WB7) $(-\hbar^2/(m+m_0))(\nabla^2)\Psi(r,t) \;=\; (E-E_0)\Psi(r,t)$

may be regarded as the "corrected" Klein – Gordon wave equation. Equation

(WB7) $(-\hbar^2/(m+m_0))(\nabla^2)\Psi(r,t) \;=\; (E-E_0)\Psi(r,t)$

was derived in 1999 by the author in a paper entitled "An Alternative Lorentz Invariant, Relativistic Wave Equation".

Chapter 9

Cosmological Analysis

Cosmological Model and General Relativity

Concepts developed in this chapter follow from most of the ideas presented by Jim Breithaupt in his book entitled "Cosmology".

Experimentally, Edwin Hubble found that the velocity of recession, v_G of a galaxy with respect to any other galaxy (not in the same group) was proportional to the distance, r_G between galaxies. Mathematically this means:

(G1) $v_G = Hr_G$

where H is the Hubble parameter (historically it was called the Hubble constant but detailed analysis shows that it is a function of time), v_G is the recessional galactic velocity, and r_G is the distance between galaxies. It will be shown that

(G2) $v_G = ((dR/dt)/R)r_G$

where R is the scale factor defined by

(G3) $r_G = Rr_0$

where r_0 is some initial separation. Note that this implies that

(G3.01) $r_0 = r_G/R$

Taking a derivative of (G3) we get

(G3.02) $v_G = r_0 dR/dt$

and plugging in from (G3.01) we get

(G3.03) $v_G = (r_G/R)(dR/dt)$

which is the same as (G2) where r_G is the distance at time t and r_0 is some initial distance at some previous time $t = t_0$. Thus, the Hubble parameter is,

(G4) $H = (dR/dt)/R$ or

(G4.1) $(dR/dt) = HR$

The Einstein tensor $E_{\sigma\mu}$ defines the curvature of spacetime and is written as

(G5) $E_{\sigma\mu} = -(8\pi G/c^2)T_{\sigma\mu}$

where $T_{\sigma\mu}$ is the stress energy momentum tensor which describes the distribution of matter (causing the curvature of spacetime), G is Newton's gravitational constant and c is the speed of light in vacuum. Einstein showed that if the universe consists of a gas with average mass density ρ, then

(G6) $(dR/dt)^2 = 8\pi G\rho R^2/3 - c^2 K + \lambda R^2/3$

where λ is Einstein's famous cosmological constant, K is the curvature of spacetime and R is the scale factor defined in equation (G3). The cosmological constant, λ was Einstein's way to force a static universe (Einstein had mistakenly assumed a static universe). Today it is known that the universe is experimentally observed to be an accelerating expansion, so set $\lambda = 0$ and equation (G6) becomes

(G7) $(dR/dt)^2 = 8\pi G\rho R^2/3 - c^2 K$

Note that by

(G4) $H = (dR/dt)/R$

equation (G7) becomes

(G7.1) $(HR)^2 = 8\pi G\rho R^2/3 - c^2 K$

and solving for K we get

(G7.2) $K = (R/c)^2(8\pi G\rho/3 - H^2)$

Define the critical density ρ_c to be

(G8) $\rho_c = 3H^2/(8\pi G)$

where H is the Hubble parameter introduced in equation (G1). This implies

(G9) $H^2 = 8\pi G\rho_c/3$

Let the density ratio, Ω be

(G10) $\Omega = \rho/\rho_c$ so that

(G11) $\rho_c = \rho/\Omega$

which implies by equation (G8) that

(G12) $\Omega H^2 = 8\pi G\rho/3$

and so equation (G7.2) becomes

(G15) $K = H^2R^2(\Omega - 1)/c^2$

Analysis shows that if $\Omega > 1$ implying $\rho > \rho_c$ then, the curvature $K > 0$, of spacetime is spherical and the universe will eventually collapse. If $\Omega = 1$ implying $\rho = \rho_c$, then the curvature $K = 0$, of spacetime is flat like an infinite plane and the universe will expand to infinity.

If $\Omega < 1$ implying $\rho < \rho_c$ then the curvature $K < 0$, of spacetime is open like a hyperbola and the universe will expand forever.

First Cause Cosmology and General Relativity

If we apply the above analysis to this first cause theory, then by (G10) $\Omega = \rho/\rho_c = 0 < 1$ since ρ, the average density of the universe as a whole is

(G16) $\rho = (\rho_1+\rho_F+\rho_2) = (m_1+m_1-2m_1)/(4/3\pi r^3) = 0$

since $m_1 = m_F$ and $m_2 = -2m_1$. The $4/3\pi r^3$ term is the spherical volume having a Schwarzschild radius r. This volume is appropriate for all three density definitions since the volume associated with ρ_1 is a sphere centered on m_1 having a Schwarzschild radius, r out to the position of m_2. The volume associated with the gravitational field density (dark matter) is centered on m_2 (the source of the field density) having a Schwarzschild radius, r out to m_1. The volume associated with dark energy, also centered on m_2 (the mass of dark energy) having a Schwarzschild radius, r out to the position of m_1.

This implies that our universe as a whole is hyperbolic since (G16) predicts the average overall density of the universe to be zero which is less than the critical density and therefore, will continue to expand forever. A way to see this is by equation

(G7) $(dR/dt)^2 = 8\pi G\rho R^2/3 - c^2 K$

and with $\rho = 0$ becomes;

(G7.1.1) $(dR/dt)^2 = -c^2 K_T$

where K_T is the total spacetime curvature. Thus,

(G7.1.2) $K_T = -(1/c^2)(dR/dt)^2 < 0$

which can be rewritten using equation (G4.1) dR/dt = HR as

(G7.1.3) $K_T = -(HR/c)^2 < 0$

which implies negative total spacetime curvature. Note that whether the positive or negative energy will expand, flatten or contract is independent of the entire universe expanding forever. Rewriting (G15) $K = H^2R^2(\Omega - 1)/c^2$ because of (G10) $\Omega = \rho/\rho_c$ yields

(G15.1) $K = (HR/c)^2(\rho/\rho_c - 1)$

where ρ is the average density of matter. Plugging in the density of equation (G16) $\rho = (\rho_1 + \rho_F + \rho_2)$, the total curvature of spacetime is

(G15.2) $K_T = (HR/c)^2[(\rho_1 + \rho_F + \rho_2)/\rho_c - 1]$

Since $\rho_F = \rho_1$ and $\rho_2 = -2\rho_1$ reduces (G15.2) to

(G15.3) $K_T = (HR/c)^2(\rho_1/\rho_c - 1) - (HR/c)^2\rho_1/\rho_c$

Thus, the total curvature of spacetime due to both positive matter and dark (matter and energy) is

(G15.4) $K_T = K_+ + K_-$

where the positive curvature of spacetime (due to m_1) is

(G15.5) $K_+ = (HR/c)^2(\rho_1/\rho_c - 1)$

and the negative curvature of spacetime due to the combined dark matter (m_F) and dark energy (m_2) is

(G15.6) $K_- = -(HR/c)^2 \rho_1/\rho_c$

Thus, three identical cases arise just as in the previous analysis of a positive mass universe (which did not take into account the dark matter or dark energy).

If $\rho_1 > \rho_c$ then, the total positive curvature $K_+ > 0$, of spacetime is spherical and the positive universe will eventually collapse.

If $\rho_1 = \rho_c$ then the total positive curvature $K_+ = 0$, of spacetime is flat like an infinite plane and the positive universe will expand to infinity.

If $\rho_1 < \rho_c$ then the total positive curvature $K_+ < 0$, of spacetime is open like a hyperbola and the positive universe will expand forever.

Figures 3, 4 and 5 in the "shape of spacetime" section of the conclusions chapter, depict K_- along with the three possibilities of K_+.

Time Dependence
of Hubble Parameter

From the definition of the Hubble parameter, in equation

(H1) $v_G = H r_G$

where r_G is the distance between any two galaxies and v_G is the velocity of one galaxy with respect to the other. Since

(H1.2) $r_G = v_G t$

where t is the elapsed time since the big bang. It is assumed that all galaxies were at the same point at time t = 0. Plugging in r_G in equation (H1.2) into equation (H1) and solving for H yields

(H1.3) $H = 1/t$

which shows how H is dependent on the time, t since the big bang. The critical density given by equation

(G8) $\rho_c = 3H^2/(8\pi G)$

can now be rewritten because of equation (H1.3) as

(H1.4) $\rho_c = [3/(8\pi G)][1/t^2]$

Density of the Positive Universe

Recall equation

(FC11.1) $m_1 = rM_P/(2L_P)$

The average density of positive mass, ρ_1 of the universe is

(D1.0) $\rho_1 = m_1/V_1$

where V_1 is the spherical volume of Schwarzschild radius r, centered on m_1, given by

(D1.1) $V = (4/3)\pi r^3$

Plugging in m_1 in equation (FC11.1) and the volume of equation (D1.1), equation (D1.0) becomes

(D1.2) $\rho_1 = 3M_P/(8\pi L_P r^2)$

Utilizing equation (D1.2) and defining the initial density ρ_{1I} when r was a Planck length L_P, yields

(D1.3) $\rho_{1I} = 3M_P/(8\pi L_P^3)$

Equation (D1.2) may be used to calculate the density of the positive matter in the universe at any time t, since by equation

(FC15.3) $r = ct + L_P$

equation (D1.2) can be rewritten as

(D1.4) $\rho_1 = [3M_P/(8\pi L_P)][1/(ct + L_P)^2]$

and since $M_P/L_P = c^2/G$, reduces it to

(D1.5) $\rho_1 = [3/(8\pi G)][1/(t + T_P)^2]$

where T_P is the Planck time. Dividing equation (D1.5) by the critical density given by (H1.4) $\rho_c = [3/(8\pi G)][1/t^2]$ and since (G10) $\Omega = \rho/\rho_c$ yields

(D1.6) $\Omega_1 = \rho_1/\rho_C = 1/(1 + T_P/t)^2$

which reveals that

(D1.6.1) lim as $t \rightarrow \infty$ of $\Omega_1 = \rho_1/\rho_C = 1$

Utilizing both equation (D1.6) $\Omega_1 = \rho_1/\rho_C = 1/(1 + T_P/t)^2$ and (H1.3) $H = 1/t$, positive and negative spacetime curvatures yield equations

(G15.5.1) $K_+ = -(R/c)^2 T_P[2t + T_P]/[t^2(t + T_P)^2]$ and

(G15.6.1) $K_- = -(R/c)^2/(t + T_P)^2 = -(R/r)^2$

Since T_P is very small, equation (G15.5.1) may be very accurately approximated by

(G15.5.2) $K_+ \approx -2(R/c)^2 T_P/t^3$ or

(G15.5.3) $K_+ \approx -2R^2 L_P/r^3$

since by (FC15.3) $r = ct + L_P \approx ct$ or

(G15.5.4) $K_+ \approx -R^2 M_P^3/(4L_P^2 m_1^3)$

since by (FC11.0) $r = 2m_1 L_P/M_P$.

Thus, the curvature of positive spacetime, K_+ (caused by the positive mass, m_1) goes from being infinitely negative ($= -\infty$) to becoming flat ($= 0$) as t, r and m_1 approach infinity. A similar analysis shows that the curvature of negative spacetime, K_- (caused by dark matter and dark energy) evolves from initially being negative to becoming flat as the universe expands to infinity. In the conclusions chapter, all the possible shapes of spacetime are depicted by figures 3, 4 and 5.

Note that at $t = 0$, the density ratio of positive density to the critical density, Ω_1 of equation (D1.6) reduces to

(D1.6.2) $\Omega_1 = \rho_1/\rho_C = 0$

and that the positive density of equation

(D1.5) $\rho_1 = [3/(8\pi G)][1/(t + T_P)^2]$

when t = 0, reduces to the initial density of equation

(D1.3) $\rho_{1I} = 3M_P/(8\pi L_P^3)$

Temperature of the Universe

Recall, that the first section of chapter 7 shows that the first cause radial solution is identical to a black hole's event radius. This means that the event horizon of the universe as a whole can be described by the first cause radius, r extending from the center of positive mass, m_1.

On page 715 of Roger Penrose's book, "The Road to Reality" (see reference section) a formula is given for the entropy of a black hole. This famous formula is known as the Bekenstein–Hawking formula and is

(T1.0) $S = kc^3 A/(4G\hbar)$

where S is the entropy, k is Boltzman's constant, c is the velocity of light in vacuum, A is the surface area of the black hole's event horizon, G is Newton's gravitational constant and $\hbar$ is Plancks constant divided by 2π. This formula may be simplified by noting that a Planck length, L_P is

Temperature Equation

(T1.1) $L_P = (G\hbar/c^3)^{1/2}$ so that (T1.0) becomes

(T2.0) $S = kA/(4L_P^2)$

The surface area A of a Schwarzschild black hole is

(T2.1) $A = 4\pi r^2$ where the event radius is given by

(FC11.0) $r = 2m_1 L_P/M_P$ and is the same as the

Schwarzschild radius given by

(FC11.5) $r = 2Gm_l/c^2$

Substituting equation (T2.1) into (T2.0) results in

(T2.0) $S = k4\pi r^2/(4L_P^2) = k\pi r^2/L_P^2$

Taking a derivative of equation (T2.0) results in

(T3.0) $dS = k2\pi r dr/L_P^2$

but by definition the infinitesimal change in entropy is given by

(T3.1) $dS = dQ/T$

where dQ is the infinitesimal energy added to the system and T is its absolute temperature. Plugging this into equation (T3.0) yields

(T4.0) $dQ/T = 2\pi k r dr/L_P^2$

However, the amount of infinitesimal energy being added to the positive universe is

(T4.1) $dQ = c^2dm_1$

and taking a derivative of equation (FC11.0) $r = 2m_1L_P/M_P$ yields

(FC11.7) $dr = 2(L_P/M_P)dm_1$ of which c^2dm_1 is

(T4.2) $c^2dm_1 = M_Pc^2dr/(2L_P) = dQ$

Plugging equation (T4.2) and equation (FC11.7) into equation (T4.0) yields

(T5.0) $M_Pc^2dr = (8\pi k/M_P)Trdm_1$

Temperature Solution

Equation (T5.0) can be rewritten in the form

(T5.1) $dr/r = (8\pi kT/(M_P^2c^2))dm_1$ whose integral is

(T5.2) $\int dr/r = (8\pi kT/(M_P^2c^2)) \int dm_1$

Carrying out the integration results in

(T6.0) $\ln r = 8\pi kTm_1/(M_P^2 c^2) + C_T$

where C_T is the integration constant to be determined. Plugging in m_1 from equation (FC11.1) $m_1 = rM_P/(2L_P)$ into equation (T6.0) yields

(T6.01) $\ln r = 4\pi kTr/(L_P M_P c^2) + C_T$

Let R_N be the radius of the universe now and T_N be its present temperature. Both of these quantities have been experimentally determined from measuring the cosmic background radiation as well as other measurements. R_N is estimated to be 13.7 billion light years (1.3 X 10^{26} meters) and T_N is 2.728 degrees Kelvin. Plugging in both T_N and R_N and solving for the integration constant, C_T yields

(T6.1) $C_T = \ln R_N - 4\pi k T_N R_N/(M_P c^2 L_P)$

Plugging equation (T6.1) into equation (T6.01) and solving for T yields

(T7.0) $T = [M_P c^2 L_P/(4\pi kr)]\ln(r/R_N) + T_N R_N/r$

This equation may be greatly simplified by noting that when $r = R_N$, then the first term to the right of the equal sign of equation (T7.0) is zero, since the natural log of $R_N/R_N = \ln(1)$ is zero, and the second term is $T_N = 2.728$ when $r = R_N$. On the other hand, when $r = L_P$, then this same first term is -1.49×10^{31}, and the second term is $T_N R_N/L_P = 2.22 \times 10^{61}$. Therefore, for all possible values of r, equation (T7.0) is very accurately approximated by

(T8.0) $T \approx T_N R_N/r$

where this first term is ignored. With equation (FC15.3) $r = ct + L_P$ the temperature can be written in terms of time since the big bang, t as

(T8.1) $T \approx T_N R_N/(ct + L_P)$

The initial temperature of the universe, T_0 is defined to be the temperature at $t = 0$, when r was a Planck length, L_P. So equation (T8.1) yields

(T8.2) $T_0 \approx T_N R_N/L_P = 2.22 \times 10^{61}$ K^0

Standard Verses First Cause Events and Temperatures

T K^0	Event	Standard Elapsed Time	First Cause Elapsed Time
$2.2*10^{61}$	Big Bang		0 sec
10^{32}	Start of Planck Era	10^{-44} sec	10^{-14} sec
10^{31}	End of Planck Era	10^{-43} sec	10^{-13} sec
$3*10^{28}$	End of Inflation Era	10^{-32} sec	$4*10^{-11}$ sec
$3*10^{12}$	Nucleon Formation	10^{-6} sec	$4*10^5$ sec
10^{10}	Helium nuclei Formation	1 sec	10^8 sec
$3*10^3$	Recombination	10^{13} sec	$4*10^{14}$ sec
30	Galaxy Formation	10^{16} sec	$4*10^{16}$ sec
2.7	Present Day	10^{17} sec	10^{17} sec

Table 9–1 Temperature Epochs

Standard Cosmological Epochs have been defined in terms of absolute temperature. Table 9–1 lists various epochs and gives a comparison between standard and first cause cosmological event times. Epochs were chosen from page 128 of Ridpath's "The Illustrated Encyclopedia of the Universe" (see reference section).

Note that the first cause theory does not contain an inflationary phase, since an accelerated expansion is built in. However, since a temperature is given at the end of the standard inflationary model, a first cause elapsed time for this event has been calculated.

Chapter 10

Summary of Equations

Laws Used For
First Cause Derivation

Physical laws or principles that were used to derive the first cause theory will now be given.

Uncertainty Principle

(U2.0) $\Delta m \Delta t \geq \hbar/(2c^2)$

(U2.1) $(\Delta m)(-\Delta t) \leq -\hbar/(2c^2)$

This is the form of the uncertainty principle which was utilized as the starting point for the first cause theory (see beginning of chapter 1).

Relativistic and Newtonian Dynamics

(N1.0.R) $F = d(m_1 v_1)/dt = Gm_1 m_2/r^2$

(N1.1.R) $-F = d(m_2 v_2)/dt = -Gm_1 m_2/r^2$

The left side of both equations (N1.0.R) and (N1.1.R) are the relativistic definitions of force originally presented by Newton's non-relativistic force, $F = ma$, where F is force, ma is the mass times its acceleration.

The right hand side of these equations express Newton's gravitational force law. However, m_1 and m_2 are functions of their own velocity. Newton was unaware that mass is a function of its velocity.

(FC1.0) $\check{r} = \check{r}_1 - \check{r}_2$

(FC1.0) is the vector definition of distance between m_1 and m_2 (see Figure 1).

(ES1.0) $m = m_0(1- (v/c)^2)^{-1/2}$

Equation (ES1.0) gives the explicit dependence of mass, m on its velocity, v. The rest mass is denoted by m_0.

(ES2.0) $v = c(1 - (m_0/m)^2)^{1/2}$

(ES2.0) is the inversion of equation (ES1.0).

(EP1.0) $E^2 - p^2c^2 = E_0^2$

Equation (EP1.0) expresses the relationship between momentum, p, dynamical energy E and rest energy E_0.

(U1.2) $E = mc^2$

(U1.3) $E_0 = m_0c^2$

(U1.2) and (U1.3) are Einstein's famous equations which relate mass, m with total dynamic energy E.

(U1.4) $p = mv$

Equation (U1.4) is the definition of momentum, p as mass, m times velocity, v.

(FC15.0) $v_{rel} = (v_1 - v_2)/(1 - v_1 v_2/c^2) = dr/dt$

(FC15.0) is the relativistic addition of velocities law.

(CM1) $\check{r}_C = (m_1\check{r}_1 - m_2\check{r}_2)/(m_1+m_2)$

(CM1) is the vector definition of the center of mass, $\check{r}_C$ for as system consisting of $m_1 > 0$ and $m_2 < 0$ whose respective positions are $\check{r}_1$ and $\check{r}_2$.

Quantum Mechanical Laws

(B7.0) $\Psi(r,t) = \Psi_0 \exp\{(rp-(E-E_0t)\}i/\hbar$

(WB7) $(-\hbar^2/(m+m_0))(\nabla^2)\Psi(r,t) = (E-E_0)\Psi(r,t)$

Equation (B7.0) express both the wave function, $\Psi(r,t)$ as a function of position, r and time, t in terms of the initial wave function, Ψ_0, momentum, p and energy, E and rest energy E_0.

Equation (WB7) is the corresponding relativistic wave equation where $\hbar$ is Planck's constant divided by 2π.

Cosmological Laws

(G1) $v_G = Hr_G$

(G1) is Hubble's famous law for the velocity of a galaxy (not in the local group), v_G where H is Hubble's parameter and r_G is the distance out to the galaxy.

(G3) $r_G = Rr_0$

(G3) expresses the galactic distance in terms of a scaling factor, R and the initial galactic distance, r_0.

(G8) $\rho_c = 3H^2/(8\pi G)$

(G8) is an expression for the critical density of matter, ρ_c in terms of Hubble's parameter and Newton's gravitational constant, G.

(G10) $\Omega = \rho/\rho_c$

Equation (G10) is the density ratio of mass density to the critical density.

(G15) $K = H^2R^2(\Omega - 1)/c^2$

(G15) is an expression for the curvature of spacetime K in terms of Hubble's parameter, the scaling factor, R and the density ratio, Ω.

The famous formula known as the Bekenstein–Hawking formula is

(T1.0) $S = kc^3A/(4G\hbar)$

where S is the entropy, k is Boltzman's constant, c is the velocity of light in vacuum, A is the surface area of the black hole's event horizon, G is Newton's gravitational constant and $\hbar$ is Plancks constant divided by 2π.

First Cause Derived Dynamical Equations

(FC4.2) $m_F = -Gm_2m_1/(rc^2)$

(FC4.2) is the definition of the mass equivalence of the repulsive gravitational field energy between m_1 and m_2.

(FC5.2) $E_T = m_1c^2 + m_2c^2 - Gm_2m_1/r = 0$

Equation (FC5.2) gives the total energy of any two masses, m_1 and m_2 along with their mutual gravitational field energy, $-Gm_2m_1/r$.

(FC6.2) $m_1c^2 + m_2c^2 + m_Fc^2 = 0$

(FC6.2) is another way of expressing the total energy of equation (FC5.2)

(FC10.5) $m_2 = -2m_1$

Equation (FC10.5) gives m_2 in terms of m_1.

(FC11.0) $r = 2m_1 L_P/M_P$

(FC11.1) $m_1 = rM_P/(2L_P)$

Equation (FC11.0) give the distance, r between m_1 and m_2 in terms of the Planck length, L_P and the Planck mass M_P. Equation (FC11.1) is the inverse of (FC11.0).

(FC11.3) $m_2 = -rM_P/L_P$

Equation (FC11.3) gives the mass of dark energy, m_2 in terms of the separation distance r. M_P and L_P are Planck mass and Planck length.

(FC11.4) $m_F = rM_P/(2L_P) = m_1$

Equation (FC11.4) expresses the fact that the equivalent field mass is the same as m_1.

(FC11.5) $r = 2Gm_1/c^2$

Equation (FC11.5) shows that the radius of the universe is the same as the Schwarzschild radius and is the same as equation (FC11.0).

(FC11.6) $m_1 = rc^2/(2G)$

Equation (FC11.6) is another way of expressing equation (FC11.1) $m_1 = rM_P/(2L_P)$.

(FC12.1) $(v_2/c)^2 = 1 - L_P^2/r^2$

(FC12.3) $(v_1/c)^2 = 1 - L_P^2/r^2$

Equations (FC12.1) and (FC12.3) express the velocity of both m_1 and m_2 in terms of separation distance, r and the Planck length, L_P.

(FC12.4) $v_1^2 = v_2^2$

(FC12.4) expresses the fact that the magnitude of m_1's velocity is the same as m_2's.

(FC13.0) $m_1 v_1 + m_F v_F - m_2 v_2 = 0 = P_T$

Equation (FC13.0) gives the total momentum, P_T in terms of m_1's momentum, m_2's momentum and the field momentum, $m_F v_F$.

(FC14.0) $F = Gr^2 (Mp/L_P)^2 / (2r^2) = c^4/(2G) = \frac{1}{2} F_P$

(FC14.0) shows that the gravitational force on m_1 due to m_2 is a constant equal to $\frac{1}{2}$ a Planck force, F_P.

(FC15.2) $t = (r - L_P)/v_{rel} = (r - L_P)/c$

Equation (FC15.2) may be taken as a definition of time in terms of separation distance, r Planck length, L_P and relative velocity $v_{rel} = c$.

(FC15.3) $r = ct + L_P$

(FC15.3) is an expression for the separation distance r, in terms of velocity of light, c , time, t and the Planck length, L_P.

First Cause Derived
Center of Mass Equations

(CM2.1) $\check{r}_1 = -\check{r}_C/3 + 2\check{r}/3$

(CM2.2) $\check{r}_2 = -\check{r}_C/3 - \check{r}/3$

(CM3) $v_C = -(v_1 + 2v_2) = v_F = v_1$

These last three equations give the vector positions of m_1, $\check{r}_1$ the vector position of m_2, $\check{r}_2$ and the center of mass velocity, v_C. The vector positions are in terms of the center of mass vector position, $\check{r}_C$ and the vector separation distance $\check{r}$.

(FCt11.0) $t = 2m_1 T_P/M_P - T_P$

(FCt11.1) $m_1 = tM_P/(2T_P) + M_P/2$

(FCt11.3) $m_2 = -tM_P/T_P - M_P$

(FCt11.5) $t = 2Gm_1/c^3 - T_P$

(FCt11.6) $m_1 = tc^3/(2G) + M_P/2$

(FCt12.1) $(v_2/c)^2 = 1 - 1/(t/T_P + 1)^2$

(FCt12.3) $(v_1/c)^2 = 1 - 1/(t/T_P + 1)^2$

Equations (FCtXX.X) have been converted from those equations originally containing the separation parameter r to ones containing the time parameter, t. Equation (FC15.3) $r = ct + L_P$ expresses the conversion.

First Cause Derived Wave Function

(B7.4.2) $\Psi_{U3}(r,t) = \Psi_{U30}(0,0)$

(B7.4.2) says that the wave function of the universe is a constant equal to $\Psi_{U30}(0,0)$.

First Cause Derived Cosmological Equations

(G7.1.3) $K_T = -(HR/c^2) < 0$

(G7.1.3) expresses how total spacetime curvature of the universe is negative.

(G15.4) $K_T = K_+ + K_-$

(G15.4) expresses the curvature of spacetime due to both positive matter and combined (dark matter dark energy).

(G15.5) $K_+ = (HR/c)^2(\rho_1/\rho_c - 1)$

(G15.5) expresses the curvature of spacetime due to m_1.

(G15.6) $K_- = -(HR/c)^2(\rho_1/\rho_c)$

(G15.6) expresses the curvature of spacetime due to both the gravitational field energy (dark matter) and dark energy m_2.

(H1.3) $H = 1/t$

(H1.3) shows how the Hubble parameter is dependent on time t, since the big bang.

(D1.2) $\rho_1 = 3M_P/(8\pi L_P r^2)$

(D1.2) expresses the density of positive matter.

(D1.3) $\rho_{II} = 3M_P/(8\pi L_P^3)$

(D1.3) expresses the initial density of positive matter.

(D1.6) $\Omega_1 = \rho_1/\rho_C = [1/(1 + T_P/t)^2]$

(D1.6) expresses the density ratio of positive density to the critical density and shows that this density ratio is essentially 1 since the Planck time, T_P is extremely small ($\sim 10^{-44}$ seconds).

(G15.5.1) $K_+ = -(R/c)^2 T_P[2t + T_P]/[t^2(t + T_P)^2]$

(G15.5.1) is the expression for the curvature of spacetime due to the positive mass of the universe which has been shown to be very accurately approximated by

(G15.5.2) $K_+ \approx -2(R/c)^2 T_P/t^3$ or in terms of r as

(G15.5.3) $K_+ \approx -2R^2 L_P/r^3$ or in terms of m_1 as

(G15.5.4) $K_+ \approx -R^2 M_P^3/(4L_P^2 m_1^3)$

(G15.6.1) $K_- = -(R/c)^2/(t + T_P)^2 = -(R/r)^2$

(G15.6.1) is the curvature of spacetime due to both dark energy and dark matter.

The Beckenstein-Hawking entropy formula

(T1.0) $S = kc^3A/(4G\hbar)$

allows for the following expression of the absolute temperature of the universe in terms of its Schwarzschild radius, r (equation (T8.0)) or elapsed time, t since the big bang (equation (T8.1))

(T8.0) $T \approx T_N R_N/r$

(T8.0) expresses the temperature of the universe in terms of its radius r, its present temperature, T_N and it present radius, R_N.

(T8.1) $T \approx T_N R_N/(ct + L_P)$

(T8.1) expresses the temperature of the universe in terms of elapsed time, t since the big bang. The speed of light in vacuum is c and L_P is a Planck length. All other variables are defined the same as for equation (T8.0) above.

Chapter 11

Tables and Charts
of Creation

Formulas used to generate all of the tables
are listed below the table unless
previously listed.

r (Lp)	t (Tp)	r1(Lp)	v1 (c)	m1(Mp)	r2(Lp)	v2 (c)	m2(Mp)	mF(Mp)
1	0	0.6667	0	0.5	-0.3333	0	-1	0.5
2	1	1.3333	0.866	1	-0.6667	-0.866	-2	1
3	2	2	0.943	1.5	-1	-0.943	-3	1.5
4	3	2.6667	0.968	2	-1.3333	-0.968	-4	2
5	4	3.3333	0.98	2.5	-1.6667	-0.98	-5	2.5
6	5	4	0.986	3	-2	-0.986	-6	3
7	6	4.6667	0.99	3.5	-2.3333	-0.99	-7	3.5
8	7	5.3333	0.992	4	-2.6667	-0.992	-8	4
9	8	6	0.994	4.5	-3	-0.994	-9	4.5

Table 0 used to generate most Charts

Explanation of Table 0

Column 1 is the quantized input (to the derived formula) of separation distance, r in terms of Planck lengths.

Column 2 is the corresponding inputs (to the derived formula) in terms of Planck times.

Column 3 is the position of m_1 (positive matter) in the center of mass system. The formula is

(CM2.1) $\check{r}_1 = -\check{r}_C/3 + 2\check{r}/3$.

$\check{r}_C = 0$ for observers in the center of mass system.

Column 4 is the velocity of m_1 with respect to the speed of light in vacuum, v_1/c in the CM. The formula is:

(FC12.3) $(v_1/c)^2 = 1 - L_P^2/r^2$

The square root of both sides is taken and solved for v_1/c.

Column 5 is the mass of positive matter m_1. The equation is

(FC11.1) $m_1 = rM_P/(2L_P)$

m_1 is calculated in units of M_P (Planck mass).

Column 6 is the position of m_2 (negative matter) in the center of mass (CM) coordinate system shown by Figure 1. The formula is

(CM2.2) $\check{r}_2 = -\check{r}_C/3 - \check{r}/3$

$\check{r}_C = 0$ for observers in the center of mass system.

Column 7 is the velocity of m_2 in the CM. The formula is:

(FC12.1) $(v_2/c)^2 = 1 - L_P^2/r^2$

The square root of both sides is taken and solved for v_2/c. c is set to 1 gives the result in term of c.

Column 8 is the mass of negative matter m_2. The equation is

(FC11.3) $m_2 = -rM_P/L_P$

m_2 is calculated in units of M_P.

Column 9 is the equivalent mass of the gravitational field energy m_F. The equation is

(FC11.4) $m_F = rM_P/(2L_P) = m_1$

The following 6 charts and associated tables all depict the physical conditions during the early big bang. Typical separation distances, r between m_1 and m_2 go out to around 9 Planck lengths. Correspondingly, typical time since creation goes out to around 9 Planck times.

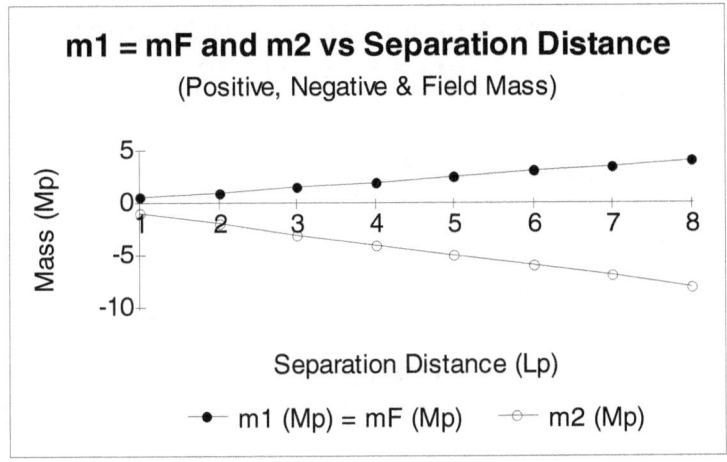

Chart 1

r (Lp)	m1 (Mp) = mF (Mp)	m2 (Mp)
1	0.5	-1
2	1	-2
3	1.5	-3
4	2	-4
5	2.5	-5
6	3	-6
7	3.5	-7
8	4	-8

Table 1 used to generate Chart 1

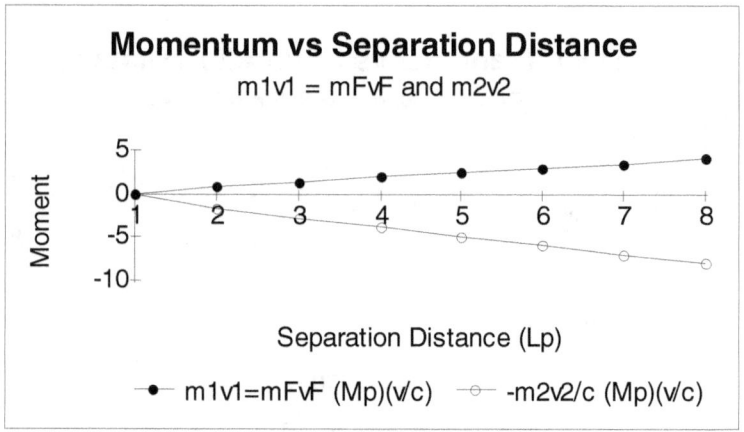

Chart 2

r (Lp)	m1v1=mFvF (Mp)(v/c)	-m2v2/c (Mp)(v/c)
1	0	0
2	0.866025404	-1.732050808
3	1.414213562	-2.828427125
4	1.936491673	-3.872983346
5	2.449489743	-4.898979486
6	2.958039892	-5.916079783
7	3.464101615	-6.92820323
8	3.968626967	-7.937253933

Table 2 used to generate Chart 2

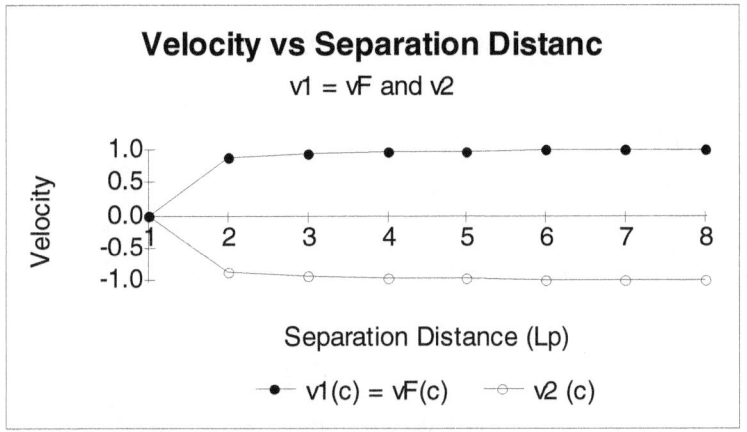

Chart 3

r (Lp)	v1 (c) = vF (c)	v2 (c)
1	0	0
2	0.866025404	-0.8660254
3	0.942809042	-0.94280904
4	0.968245837	-0.96824584
5	0.979795897	-0.9797959
6	0.986013297	-0.9860133
7	0.989743319	-0.98974332
8	0.992156742	-0.99215674

Table 3 used to generate Chart 3

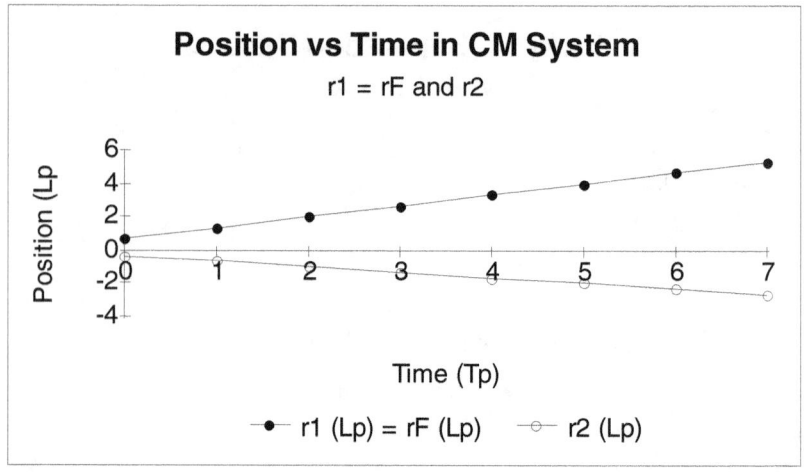

Chart 4

r (Lp)	t (Tp)	r1 (Lp) = rF (Lp)
1	0	0.666666667
2	1	1.333333333
3	2	2
4	3	2.666666667
5	4	3.333333333
6	5	4
7	6	4.666666667

Table 4 used to generate Chart 4

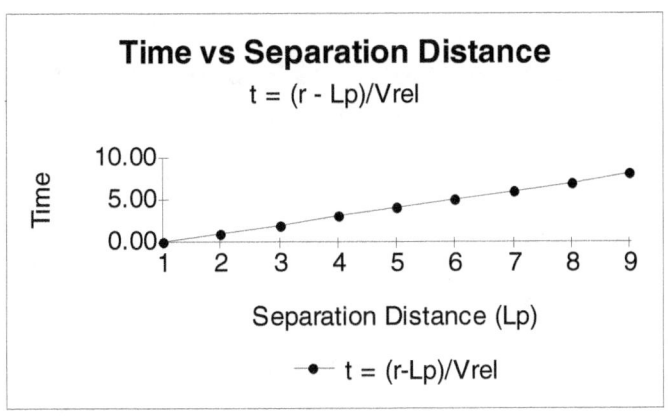

Chart 5

r (Lp)	t = (r-Lp)/Vrel	Vrel = (v1-v2)/(1-(v1v2/c*c))
1	0.00	0
2	1.01	0.989743319
3	2.00	0.998268397
4	3.00	0.999479573
5	4.00	0.999791732
6	5.00	0.999900808
7	6.00	0.999946858
8	7.00	0.999968999

Table 5 used to generate Chart 5

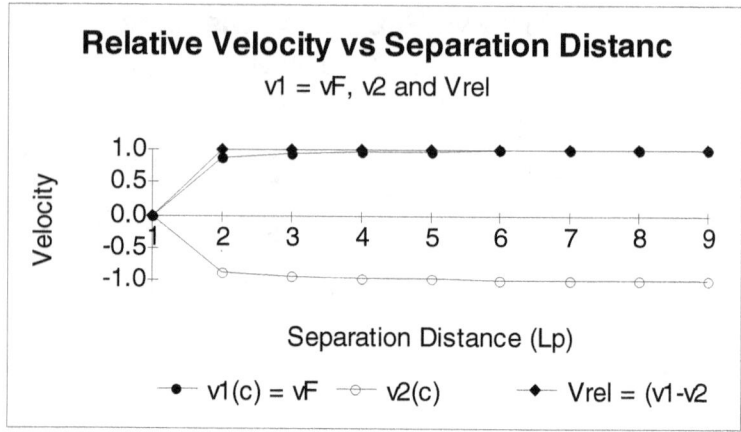

Chart 6

r (Lp)	v1(c) = vF(c)	v2(c)
1	0	0
2	0.866025404	-0.866025404
3	0.942809042	-0.942809042
4	0.968245837	-0.968245837
5	0.979795897	-0.979795897
6	0.986013297	-0.986013297
7	0.989743319	-0.989743319
8	0.992156742	-0.992156742
9	0.99380799	-0.99380799

Table 6 used to generate Chart 6

Discussion of Charts and Tables

From the charts, both positive energy (matter and field) and negative (dark) energy are rapidly increasing in such a way that the total energy is a balanced constant. Essentially, these charts depict the details of early inflation.

For instance, from table 1, chart 1, table 2 and chart 2, both the mass and momentum of the positive and negative systems are linearly increasing. Table 1 and Chart 1 show that the universe unfolded itself in discrete units of separation distance (L_P) and in discrete units of time (T_P). Positive mass energy increased in integral multiples of ½ Planck mass. Negative mass energy increased in integral multiples of –1 Planck mass. Note that the positive repulsive gravitational field energy being equal to the positive mass energy also increased in integral multiples of ½ Planck mass.

Table 3 and chart 3 shows that the velocity of the center of mass of the positive energy as well as the speed of the field energy are quickly approaching the speed of light. The velocity of the center of mass of negative energy is quickly approaching light speed in a direction opposite to the positive energy velocity. Even though positive energy, dark energy and field energy will undergo many transformations, they are balanced energy wise. Positive energy and negative dark energy can be represented by only 2 points (their respective centers of mass) which accelerate away from each other. However, because positive energy and dark energy are unbalanced toward the negative, it is the positive explosive field energy filling the spacetime between them that must account for their difference in energy and momentum (see Figure 2 in the last chapter). Chart 3 is the first cause version of cosmological inflation.

Table 4 and chart 4 shows the positions of both positive and dark energy in their center of mass system of coordinates.

Table 5 and chart 5 shows that if time is defined to be the two system's separation distance divided by their relative velocity (using the relativistic addition of velocities law). What this means is that in all charts, the separation distance may also be represented by the speed of light multiplied by the time plus a Planck length. Separation distance is measured in Planck lengths (L_P) while time is measured in Planck times (T_P). If both separation distance and time are linear, then spacetime is linear.

Table 6 and chart 6 shows the relative velocity of both the positive energy system and the dark (negative) energy system as a function of separation distance (r).

Chapter 12

Conclusions

Big Bang Triggering Mechanism

Our universe was born from a pair of quantum mechanical vacuum fluctuation which during a negative and positive Planck time produced the equivalent of a negative and positive energy Planck mass pair, separated by one Planck length. The system conserves energy and appears as three initial primordial systems. These are the positive mass, positive repulsive gravitational field energy and negative mass. This was enough for a repulsive gravitational force to evolve and violently separate positive matter from negative matter.

Thus, the creation scenario began when time fluctuated backwards by $-1T_P$, and a negative mass m_2 appears having mass, $-1M_P$ with the positive repulsive field energy equivalent having $+\frac{1}{2}M_P$. Time then moves forward by $+1T_P$ returning back to zero, $(t = 0)$, and a positive mass m_1 appears having a mass of $\frac{1}{2}M_P$. Thus at time zero, m_1 and m_2 appear and are separated by one Planck length (L_P) with a net mass of $-\frac{1}{2}M_P$. The positive repulsive field energy filling the spacetime between m_1 and m_2 simultaneously appears with an energy equivalent mass of $+\frac{1}{2}M_P$ to conserve energy.

Explanation of the Accelerated Expansion and Possible Higgs Field

With another forward tick of the Planck clock, and a separation distance of 2 Planck lengths, both the positive and negative energy have increased to double their rest energies (as function of their velocities) and have been accelerated away from each other with m_1's velocity at .867c and m_2's velocity of −.867c. The gravitational field energy mass equivalent has also doubled. The process conserves both energy and momentum and the run away expansion continues with both m_1 and m_2 moving under the influence of a constant Planck force of $\frac{1}{2}F_P$. Note that both the positive energy and dark energy systems continue to accelerate away from each other. Each system approach relative light speed very quickly because of this tremendous repulsive gravitational field. This field may be an alternate representation of the Higgs Field.

As the positive and negative masses separate at relative light speed, their centers of mass form a gravitational dipole whose increasing poles of $+m$ and $−2m$ and gravitational field strength equivalent of $+m$.

Candidate for Missing Dark (Negative) Energy

On page 49 of the January 2001 issue of Scientific American, J. P. Ostriker and P. J.

Steinhardt in their quintessential universe, present a "recipe for the universe" as a pie chart showing the percentages of dark energy, exotic dark matter, ordinary nonluminous matter, ordinary visible matter and radiation. This is shown in Table 12–1, Row 1. Note that only the

Dark Energy	Exotic Dark Matter	Non luminous Matter	Visible Matter	Radiation	Theory
70	26	3.5	0.5	0.005	Quintessence
66.67	28.88	3.89	0.555	0.005	First Cause

Table 12–1 Energy and Matter Percentages

dark energy is negative. All the other forms of energy are positive. Since, this (first cause) theory says that the dark energy must have twice the magnitude of the positive energy, it follows that if the magnitude of all the positive energy is 33.33 % , then the magnitude of all the dark energy must be 66.67 %. In other words, if the positive energy is represented by x, then the dark energy must be 2x. Since 2x (dark energy) + x (positive energy) = 3x = 100 %. If one solves this equation for x, one finds that x must be 33.33 % and the dark energy must be 66.67 %. No other percentages are possible. Of course, quintesential theory is not aware of first cause's balancing repulsive gravitational field energy that pervades the spacetime between positive matter and negative dark energy.

Notice that in the quintessential theory, all the positive energy adds up to just over 30%. Thus, only a slight modification of its percentages would make the quintessence theory compatible with the (first cause) theory. These interpolated values are shown in Table 12–1, Row 2.

Negative Binding Energy

Negative mass particles have not been directly observed in the lab. However, negative energy is well known in connection with energy systems which are bound to other energy systems. For instance the electron proton (hydrogen atom) binding energy is negative. It takes positive energy added to the electron proton system in order to separate the electron from the proton. The electrical potential energy between unlike charges is negative (attractive). The electrical potential energy between like charges is positive (repulsive). The opposite is true of gravitational potential energy. By definition, the gravitational potential energy of any two positive masses is negative (attractive). Experimentally, it is known that this gravitational potential energy is proportional to the product of both masses. Thus, it can easily be shown that the gravitational potential energy between any two negative masses would also be negative (attractive) and the gravitational potential energy between a positive mass and a negative mass would be positive (repulsive). It could be that Ostriker and Steinhardt's quintessence field is equivalent to a gravitational repulsive field which exists in the spacetime between the positive energy and dark energy systems of the universe.

Are We Positive or Negative Energy?

It must be mentioned that it appears that we live in the positive energy world since we measure dark energy to be around 67% of the total. If we had measured the dark energy to be 33% of the total, then we could assume that we live in the negative energy world. Nevertheless, dark energy and positive energy represent entangled parallel worlds.

Invisible Dark (Negative) Energy

Note that negative photons would repel and be repelled by positive photons. Thus, no negative photons will ever arrive from the negative energy spacetime. The negative part of the universe is invisible to the positive. Likewise the positive energy system is invisible to the negative energy system. This may explain why negative energy particles have never appeared in the labs.

Solution to the Asymmetry Problem

If it is assumed that negative energy is composed of negative anti-matter and positive field energy is positive matter, then matter and anti-matter are exactly balanced. This explains the asymmetry that our positive energy world contains much more matter than anti-matter. Thus, in the negative energy world, negative anti-matter would rule with its negative anti-positrons encircling negative anti-atoms with negative anti-nuclei composed of negative anti-quark trios. Thus, the dark energy (negative mass energy) plus the positive energy (positive mass energy and positive gravitational field energy) conserve matter-anti-matter symmetry (matter-anti-matter = 0). Thus, all the repulsive gravitational field energy (pushing apart the positive and negative energy systems) would be composed of positive matter, so as not to violate matter anti-matter symmetry. In dark energy spacetime, if a negative mass electron appears, it would quickly be annihilated by the negative mass anti-positron with the result of either 2 or 3 dark (negative) photons.

Moreover, black holes exist in positive energy spacetime, and thus, negative black holes (having negative mass) may exist in negative energy spacetime.

Thus, first cause theory shows the existence of growing parallel worlds, one with positive matter and the other with negative anti-matter entangled by their mutual positive repulsive gravitational field energy. Both of these worlds are seen to have common parents in the original creation of a positive and negative Planck mass pair.

Candidate for Missing Dark Matter

Part of the so called missing dark matter of the universe may exist in the spacetime between galactic masses. This theory predicts that a positive repulsive gravitational field energy permeates all of spacetime. This gravitational field may be identified with the Higgs field of the standard model. Recall that this field energy (that pushes against the positive and negative energy) is equivalent to all the positive mass energy in the universe. Moreover, since it is this positive gravitational field energy that is causing all the galaxies to accelerate, this acceleration causes the mass of all the galaxies to increase with respect to time. This increase potentially offers an explanation of why stars are observed to behave as if their galaxy is more massive than their estimated observational mass.

If the Higgs Field gives mass to elementary particles, it also must be responsible for the observed rest mass energy (m_0c^2) of each particle. From the definitions of Planck mass, M_P and Planck length, L_P it is easy to show that

(HG1) $c^2 = GM_P/L_P$

where G is Newton's gravitational constant and therefore

(HG1.1) $m_0c^2 = Gm_0M_P/L_P$

which is the gravitational potential energy between any rest mass, m_0 and a negative Planck mass, $-M_P$, separated by a distance of one Planck length L_P. This means that the total rest mass energy of any particle is equivalent to how much energy would be required to bring that mass to rest within a Planck length of a negative Planck mass (against its repulsive gravitational field). Thus, any rest mass energy is seen to be equivalent to the gravitational potential energy between that mass at rest and a negative Planck mass at rest, separated by a distance of one Planck length. This also may be another way of looking at how the Higgs field causes rest mass energy.

Recall that the creation condition of this theory, is that at the instant of the big bang, a negative Planck mass, separated by a Planck length from a ½ positive Plank mass, along with their mutual positive repulsive gravitational field energy equivalent of a ½ positive Planck mass at $t = 0$, all appear simultaneously at rest out of the vacuum.

First Cause and The Steady State Theory

This first cause theory also contains many of the ideas of the steady state theory of the universe put forth by Fred Hoyle, Hermann Bondi and Thomas

Gold. This theory was based upon the idea that matter is continuously created in the voids between galaxies as they move apart. The first cause theory also contain this idea in the sense that the positive mass of the universe increases as the universe expands. In fact, using equation (FCt11.6) $m_1 = tc^3/(2G) + \frac{1}{2}M_P$, the total mass of the positive universe including the repulsive field energy mass equivalent is $2m_1 = tc^3/G + M_P$. Therefore, the rate of increase of the positive mass of the universe is $2dm_1/dt = c^3/G$ and correspondingly, the rate of increase in the negative mass is $dm_2/dt = -c^3/G$. This means that a steady stream of positive mass and negative mass is flowing into our universe.

In a book entitled "The Illustrated Encyclopedia of the Universe" (ISBN 0–8230–2512–8), Watson–Guptill Publications/New York, 2001 on page 142, it mentions that dark matter is thought to make up 90% of the universe. The first cause theory predicts it is 100% since the positive gravitational field energy of the universe is equivalent to the total positive mass of the universe. Recall, that it is this gravitational field energy that is causing the galaxies to accelerate away from each other. Galaxies undergoing relative acceleration constantly gain mass with respect to each other. This means that galactic masses have been underestimated if it has been assumed that galaxies move away at constant rates (unaccelerated). In other words, mass is constant at constant speed, but mass increases with acceleration. However, because of equation

(ES1.2) $dv/dt = ((c^2 - v^2)/mv)dm/dt$

the acceleration

(ES1.2.1) $dv/dt \rightarrow 0$ in the limit as $v \rightarrow c$

Even though the acceleration of galaxies approach zero, the rate at which the positive universe gains mass and the rate of increase of repulsive gravitational equivalent field energy is constant and given by

(FCt11.6.1) $dm_l/dt = \frac{1}{2}c^3/G = dm_F/dt$

The Seeds of Positive Mass

It is estimated that $\frac{1}{2}$ Planck mass $(1.09 \times 10^{-8}$ Kilograms) contains the equivalent mass of (6.55×10^{18}) hydrogen atoms. If a volume of spacetime that contained these atoms (or perhaps other forms of primordial energy at absolute zero temperature) were suddenly subjected to an unbalanced push of $\frac{1}{2}F_P$ ($\frac{1}{2}$ a Planck force), a temperature of some 10^{61} K^0 would have resulted. Thus, a hemispherical Gaussian distribution (in energy and spacetime) of escaping   particles (including zero rest mass particles such as photons and neutrinos) moving at the fastest possible speed, would have resulted. Thus, this initial positive energy distribution of particles were destined to be the seeds of the positive universe. Since the temperature of the universe immediately before the creation event, was absolute zero, it is also assumed that the initial $\frac{1}{2}$ Planck mass consisted of a compact spherical distribution of hydrogen (protons & electrons).

As the temperature dropped, and the universe expanded and cooled, it is assumed to have gone through all the standard cosmological epochs defined by table 9–1. These included nucleon formation at 3×10^{12} K^0, fusion (helium production) at 10^{10} K^0, recombination (atom formation) at 3000 K^0 and finally us at 2.7 K^0.

The Evolution
of the Positive Universe

Using equations (FC11.0) $r = 2m_1L_P/M_P$ and (FC15.2) $t = (r - L_P)/c$, if the positive mass of the universe was equivalent to the mass of our milky way galaxy or 4×10^{41} kilograms, then the radius (r) of the universe was 5.94×10^{14} meters. By (FC15.2), this translates to about 1.98×10^6 seconds or 550 hours or 22.92 days. Thus, this theory predicts that the mass of the positive universe was equivalent to the mass of our milky way galaxy in less than 23 days. Similarly, it is also estimated that the mass of the positive universe was

1 solar mass (2 X 10^{30} kilograms) in about 8 microseconds.

Big Early Galaxies

In the July 2004 issue of Scientific American on page 32, George Musser discusses the discovery of young massive galaxies that seem to have existed before their time. For example, galaxy J02399 had 300 billion solar masses when only 2.4 billion years old! This could be explained by calculations of the first cause theory similar to the discussion above, entitled "Predicting the Evolution of the Positive Universe". In fact, at 2.4 billion years, by (FCt11.1) $m_1 = tM_P/(2T_P)+M_P/2$, the mass of the positive energy (m_1) comes out to around 7.65 X 10^{21} Solar masses. Thus, this theory says that there was more than 25 billion equivalent J02399 galaxy masses available at this time.

First cause theory says that the amount of positive energy in the universe grows in proportion to its radius (the distance (r) between the positive (m_1) and negative (m_2) system) which is equal to light speed times time (ct) plus a Planck length ($r = ct + L_P$).

Cosmological Predictions

Using equation (FC11.0) $m_1 = rM_P/(2L_P)$, if the radius (r) of the universe, is $R_U = 13.7 \times 10^9$ light years $= 1.296 \times 10^{26}$ meters, then the positive mass of the universe is $M_{+U} = 8.7 \times 10^{52}$ kilograms with the negative mass of the universe being $M_{-U} = -17.4 \times 10^{52}$ kilograms. The total mass of the positive and negative system is $M_{U2} = M_{+U} + M_{-U} = -8.7 \times 10^{52}$ kilograms. To balance the energy, the positive repulsive gravitational field energy equivalent mass is also $M_{UGF} = 8.7 \times 10^{52}$ kilograms.

Assuming all the positive mass of the universe is somehow spatially partitioned and distributed among the galaxies and/or galaxy like clusters and if the mass of the milky way galaxy is $M_{MW} = 4 \times 10^{41}$ Kg, then the number of equivalent milky way galaxies is $N_G = M_{+U}/M_{MW} = 2.175 \times 10^{11}$ galaxies.

Similarly, if the sun's mass is $M_S = 2 \times 10^{30}$ Kg, the number of equivalent suns would be $N_S =$

$M_{+U}/M_S = 4.35$ X 10^{22} suns. Therefore, the number
of equivalent suns per equivalent milky way galaxy
is $N_S/N_G = 2$ X 10^{11} stars per galaxy.

After the Big Bang

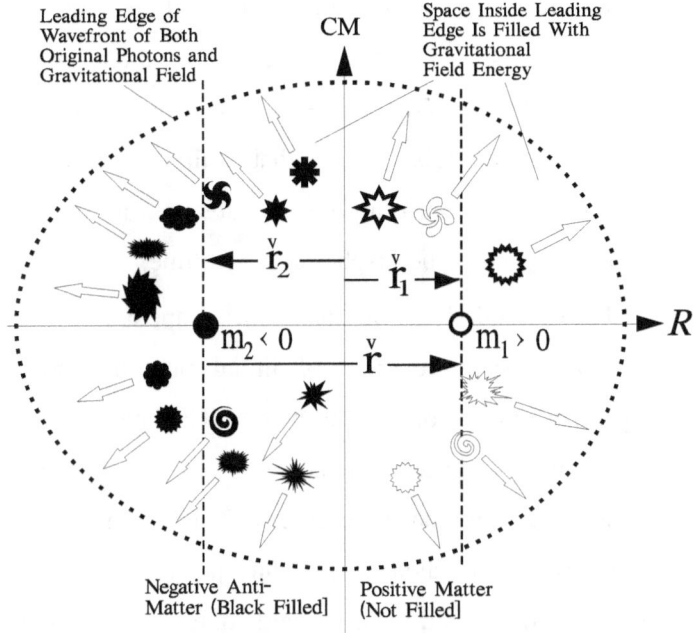

Figure 2 Center of Mass (CM)
Coordinate System
After the Big Bang

Figure 2 is a representation of our universe after
the big bang. The arrows show the direction of

motion of only a few depicted galaxies. Dark energy is represented by the black filled symbols. Positive energy is represented by unfilled symbols. Keep in mind, that in the charts, as well as Figure 2, the separation distance is the distance between the center of mass of positive matter, m_1 and the center of mass of negative (dark energy) anti-matter, m_2. The two dotted vertical lines passs through these two centers of mass. Note that both matter and negative anti-matter (dark energy) are each moving into an expanding hemispherical spacetime.

The dotted line of the oval shaped figure represent the wave front of initial photons and repulsive gravitational field energy. The spacetime inside the oval between galaxies is filled with this positive gravitational field energy. This energy has a field mass equivalent to m_1. Thus, positive matter, negative matter and positive field energy sum up to zero.

Shape of Spacetime

Recall that if the average density of positive mass, ρ_1 is greater than, equal to, or less than the critical density, ρ_c then positive spacetime curvature is either spherical (figure 3), flat (figure 4), or open (figure 5) respectively. In all of these figures, only the initial negative spacetime curvature is shown. Equation

(G15.4) $K_T = K_+ + K_-$

shows that the total curvature of the universe is the sum of a positive curvature, K_+ (due to m_1) and a negative curvature, K_- (due to both dark energy m_2 and positive field energy or dark matter, m_F). Equation

(G15.5) $K_+ = (HR/c)^2(\rho_1/\rho_c - 1)$

represents the positive spacetime curvature and equation

(G15.6) $K_- = - (HR/c)^2(\rho_1/\rho_c)$

is the negative spacetime curvature.

Note that if K_+ had been positive, then Figure 3 would depict positive spacetime curvature.

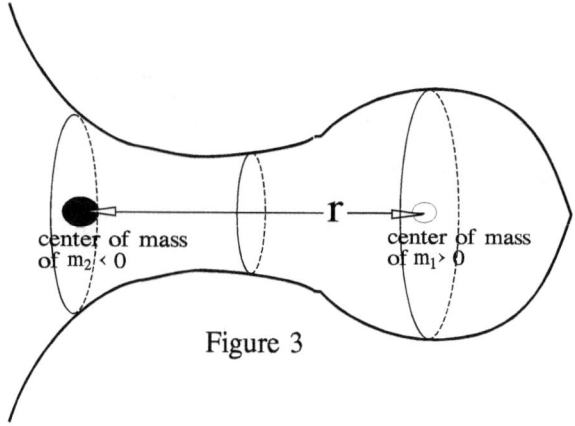

center of mass
of $m_2 < 0$

r

center of mass
of $m_1 > 0$

Figure 3

and would have meant that positive spacetime would be spherical and the positive component of the universe would have eventually collapsed into a positive big crunch. Figure 4 depicts flat spacetime as K_+ approaches zero.

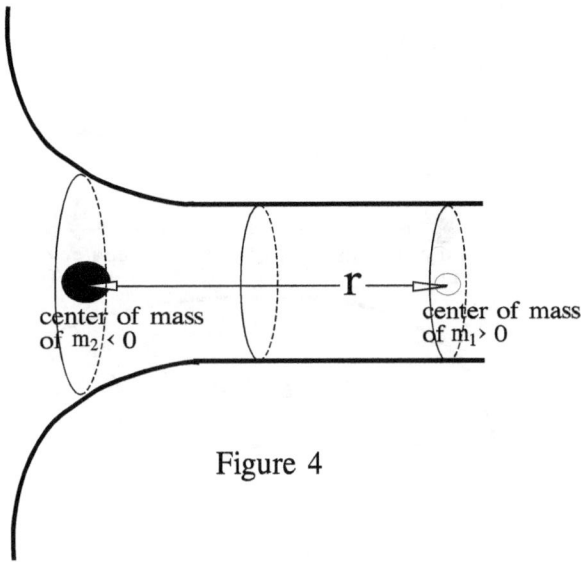

center of mass
of $m_2 < 0$

r

center of mass
of $m_1 > 0$

Figure 4

This depicts the zero curvature of positive spacetime as the universe expands to infinity.

Figure 5 depicts spacetime curvature when both K_+ and K_- were initially less than zero.

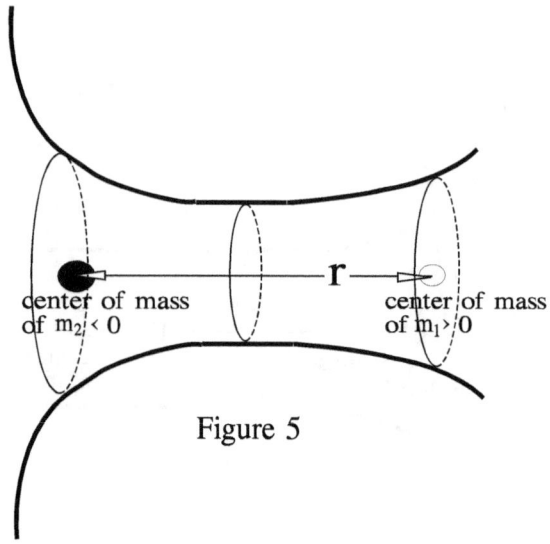

center of mass of $m_2 < 0$

r

center of mass of $m_1 > 0$

Figure 5

Flatness Problem Solved

The equation for the ratio of density, ρ_1 to the critical density, ρ_C is

(D1.6) $\Omega_1 = \rho_1/\rho_C = [1/(1 + T_P/t)^2]$

This density ratio quickly converges to a value of 1 (unity). Thus, the curvatures of spacetime are given by equations

(G15.5.1) $K_+ = -(R/c)^2 T_P[2t + T_P]/[t^2(t + T_P)^2]$ and

(G15.6.1) $K_- = -(R/c)^2/(t + T_P)^2 = -(R/r)^2$

(G15.5.2) $K_+ \approx -2(R/c)^2 T_P/t^3$ or in terms of r as

(G15.5.3) $K_+ \approx -2R^2 L_P/r^3$ or in terms of m_1 as

(G15.5.4) $K_+ \approx -R^2 M_P{}^3/(4L_P{}^2 m_1{}^3)$

which all approach zero as t, r and m_1 approach infinity. The curvature of positive spacetime due to positive matter evolves from being infinitely negative (figure 5) to eventually becoming flat (figure 4). Negative spacetime curvature, K_- also evolves from initially being negative to becoming flat at infinity. Note that these results are achieved without having to invoke the inflationary model.

More Studies

A study could be done to determine a general expression for the relative velocities, distances, spectrum shifts and clocks with respect to the observed recession of galaxies using the first cause model. It was noticed that all triangles formed from the position of the positive energy system's center of mass, the position of an arbitrary galaxy and the position of the center of mass of the negative energy system have a common leg equal to $r = ct + L_P$ (the radius of the universe).

General relativity could be examined to see how the equations behave when the energy of a massive body is negative since it already covers the zero and positive case. There should be general relativistic relationships which could clarify the behavior of spacetime under the matter and negative anti-matter first cause scenario.

Studies could be done on what impact the first cause scenario would have on superstring theory, twistor theory and loop quantum gravity.

GLOSSARY

Electrical Symmetry: Any two charges regardless of their signs (positive or negative) separated by a finite distance apart, released from rest, always electrically interact to cause the two charges to move in opposite directions. They both develop opposite directed velocities as well as opposite directed accelerations.

Gravitational Symmetry: Any two masses regardless of their signs (positive or negative) separated by a finite distance apart, released from rest, always gravitationally interact to cause the two masses to move in opposite directions. They both develop opposite directed velocities as well as opposite directed accelerations.

Material Energy: Matter or anti-matter stored energy associated with mass, in contrast to field energy which permeates spacetime.

Newton's Universal Gravitational Constant: $G = 6.67 \times 10^{-11}$ newton-meters^2/kilograms^2

Planck Acceleration: $A_P = c/T_P = 5.56 \times 10^{35}$ meters/second^2

Planck Boson Spin: $\hbar = 1.0552 \times 10^{-34}$ joule–seconds

Planck Clock: Each tick is a Planck Time. The smallest event time.

Planck's Constant: $h = 6.63 \times 10^{-34}$ joule-seconds

(Planck's Constant)/2π: $h/2\pi = \hbar = 1.0552 \times 10^{-34}$ joule-seconds

Planck Energy: $E_P = M_P c^2 = 1.96 \times 10^{9}$ joules

Planck Fermion Spin: $\hbar/2 = 5.276 \times 10^{-35}$ joule–seconds

Planck Force: $F_P = GM_P^2/L_P^2 = M_Pc/T_P = c^4/G =$ 1.21×10^{44} newtons

Planck Frequency: $W_P = M_Pc^2/\hbar = 1.857 \times 10^{44}$ hertz

Planck Length: $L_P = (\hbar G/c^3)^{1/2} = 1.616 \times 10^{-35}$ meters

Planck Mass: $M_P = (\hbar c/G)^{1/2} = 2.177 \times 10^{-8}$ kilograms

Planck Mass Pair = a positive ½ Planck mass, $½M_P/2$, and a negative ½ Planck mass, $- ½M_P/2$. However the negative ½ Planck mass is equivalent to a negative Planck mass, $-M_P$ and an associated repulsive gravitational field energy equivalent of a positive ½ Planck mass, $+ ½M_P/2$.

Planck Momentum: $P_P = M_Pc = 6.53$ kilogram–meters/second

Planck Time: $T_P = (\hbar G/c^5)^{1/2} = 5.391 \text{ X } 10^{-44}$ seconds

Radius of the Universe: Distance between the center of mass of all positive energy and the center of mass of all negative energy. Current estimate ~ 13.7 Billion Light Years = $1.296 \text{ X } 10^{26}$ meters.

Speed of Light in Vacuum: $c = 3.00 \text{ X } 10^8$ meters/second

Vacuum Symmetry Breaking: Positive matter and positive field energy balanced by dynamic negative anti-matter.

Vacuum Symmetry: No net positive or negative material energy with no net motion, (equal positive and negative time fluctuations) and net empty spacetime. Population averages for energy, momentum, angular momentum and charge were zero. The perfect harmony of nothingness or emptiness. The void before the big bang.

Fundamental Physical Laws

Preliminary Definitions

I. **Bold** mathematical single letters refer to vectors.

II. The symbol, $i = (-1)^{1/2}$ always occurs as the fourth (time component) of all Einsteinian four dimensional vectors.

III. The symbol, c is the speed of light in vacuum.

1. **Position of an energy system:** Referring to Figure 2, a normal Cartesian coordinate system shows the x,y,z position of the system S at time t.

Newtonian: $\mathbf{r}_N = (x,y,z)$

Einsteinian: $\mathbf{r}_E = (x,y,z,ict)$

2. **Velocity of an energy system:** At time t_2, the position of the system was at position 2 (r_2).

Initially at time t_1, the position of the system was at position 1 (r_1). The average velocity of the system is the distance traversed by the system in moving from position 1 to position 2 ($r_2 - r_1$) divided by the time it took for the system to move between the two positions ($t_2 - t_1$). The direction of the velocity is from position 1 to position 2. The instantaneous velocity **v** is realized by letting t_2 approach t_1. Mathematically, the instantaneous velocity of a system is a vector quantity.

$\mathbf{v} = \lim$ as $t_2 \rightarrow t_1$ of $[(\mathbf{r_2} - \mathbf{r_1})/(t_2 - t_1)]$ or

$\mathbf{v} = d\mathbf{r}/dt$

Newtonian: $\mathbf{v}_N = (v_x, v_y, v_z)$

Einsteinian: $\mathbf{v}_E = (v_x, v_y, v_z, ic)$

3. **Acceleration of an energy system:** At time t_2, the velocity of the system was (v_2). Initially at time t_1, the velocity of the system was (v_1). The average acceleration of the system is the change in the

velocity of the system in going from v_1 to v_2, $(v_2 - v_1)$ divided by the time it took for the system to go from v_1 to v_2, $(t_2 - t_1)$. The direction of the acceleration is from v_1 to v_2. The instantaneous acceleration $\mathbf{a}$ is realized by letting t_2 approach t_1. Mathematically, the acceleration of a system is a vector quantity.

$\mathbf{a} = \lim$ as $t_2 \to t_1$ of $[(\mathbf{v_2} - \mathbf{v_1})/(t_2 - t_1)]$ or

$\mathbf{a} = dv/dt$

Newtonian: $\mathbf{a}_N = (a_x, a_y, a_z)$

Einsteinian: $\mathbf{a}_E = (a_x, a_y, a_z, 0)$

4. **Momentum of an energy system:** The product of the system's mass and its velocity $\mathbf{v}$ is called its momentum and denoted by $\mathbf{p}$. It is a vector quantity having direction $\mathbf{v}$.

$\mathbf{p} = m\mathbf{v}$

Newtonian: $\mathbf{p}_N = (p_x, p_y, p_z)$

Einsteinian: $\mathbf{p}_E = (p_x, p_y, p_z, iE/c)$

where E is the total energy $= mc^2$.

5. **Force on an energy system:** The instantaneous rate of change of a system's momentum with respect to time. Its definition is similar to the definition of velocity. At time t_2 it has momentum 2. At time t_1, it had momentum 1. The average force is the difference in momentum (momentum 2 – momentum 1) divided by the time difference $(t_2 - t_1)$. The instantaneous rate is realized when t_2 approaches t_1.

$\mathbf{F} = \lim$ as $t_2 \to t_1$ of $[((m\mathbf{v})_2 - (m\mathbf{v})_1)/(t_2 - t_1)]$ or

$\mathbf{F} = d(m\mathbf{v})/dt$

Newtonian: $\mathbf{F}_N = m_0 d\mathbf{v}/dt = m_0\mathbf{a}$
where m_0 is the rest mass and $\mathbf{a}$ is the acceleration.

Einsteinian: $\mathbf{F}_E = d(m\mathbf{v})/dt = md\mathbf{v}/dt + \mathbf{v}dm/dt$

$$= m_0\mathbf{a}(1 - (v/c)^2)^{-3/2} = (c^2/v)dm/dt$$

where $\mathbf{v}$ is m's velocity and $\mathbf{a}$ is m's acceleration.

6. **Mass density:** Mass m, per unit volume V.

Average mass density = ρ_{mavg} = m/V

Instantaneous mass density = ρ_m = dm/dV

7. **Pressure on a surface**: The applied force F, per unit surface area, A.

Average pressure = P_{avg} = F/A

Instantaneous pressure = P = dF/dA

8. **Angular momentum of an energy system:** Let the vector from the origin to the position of the system be called the position vector ($\mathbf{r}$). The angular momentum of the system ($\mathbf{L}$) is then the

ordinary vector cross product (**x**), of the position vector with the system's momentum vector (m**v**).

$$\mathbf{L} = \mathbf{r} \text{ x } m\mathbf{v}$$

9. **Charge density:** Amount of charge q, per unit volume V.

Average charge density $= \rho_{qavg} = q/V$

Instantaneous charge density $= \rho_q = dq/dV$

10. **Electrical current:** The instantaneous change of charge q with respect to time.

$i = \lim$ as $t_2 \rightarrow t_1$ of $[(q_2 - q_1)/(t_2 - t_1)]$ or

$i = dq/dt$

11. **Electrical current density:** The electrical current i per unit cross sectional area A of conductor. The unit vector **u** has a direction of the

current i along the conductor perpendicular to the cross sectional area.

$$\mathbf{J}_i = \mathbf{u}i/A \text{ where } i = dq/dt$$

In a conductor with conductivity σ_c, the current is in the direction of the electric field $\mathbf{E}$ and the electrical current density is the product of the conductivity and electric field.

$$\mathbf{J}_i = \sigma_c\mathbf{E}$$

12. **Mole**: The mass of Avogadro's number of identical molecules or Avogadro's number of identical atoms expressed in grams. One mole of molecules is the molecular weight of the molecule expressed in grams. One mole of atoms is the atomic weight of the atom expressed in grams.

Mechanical Laws

1. Newton's Laws of Motion:

1.1   A body will remain at rest or in motion at a constant velocity unless acted on by an unbalanced external force.

1.2   The force on a body is proportional to its acceleration and the constant of proportionality is the rest mass (when the body is at rest), m_0 of the body.

$$\mathbf{F} = m_0\mathbf{a}$$

(Newton was unaware that mass is a function of its velocity.)

1.3   The force of one body on a second body is equal and opposite to the force of the second body on the first body or for every action, there is an equal and opposite reaction.

$\mathbf{F}_{12} = -\mathbf{F}_{21}$

1.4   Newton's Universal Law of Gravitation says that any two energy systems having mass attract each other with a force ($\mathbf{F}$) proportional to the product of their masses m_1 and m_2 and inversely proportional to the square of the distance (r) between their mass centers. The force is in a direction between the centers of m_1 and m_2, causing them to attract one another and is denoted by the unit vector $\mathbf{r}_u$. G is the constant of proportionality known as Newton's Gravitational Constant. This force is

$\mathbf{F} = \mathbf{r}_u Gm_1m_2/r^2$

deriving the gravitational potential energy, V between m_1 and m_2 as

$V = -Gm_1m_2/r$

Newtonian: Mass is the cause of the gravitational field.

Einsteinian: Mass energy and momentum warp four dimensional spacetime into a gravitational field.

2. **Quantum Mechanical Laws**:

2.1 An energy system may be described by a wave function. The total energy operator $\hat{H}$ (known as the Hamiltonian) operating on the wave function (Ψ) yields the total energy eigenvalue (E) of the system represented by the wave function. Energy eigenvalues (E) are the allowable energy states that the system may assume. Similarly, other operators operating on the wave function yield other information (such as the spin, momentum, angular momentum, etc.) about the system.

$$\hat{H}\Psi = E\Psi$$

2.2 The square of the wave function $\Psi^{*}\Psi$, multiplied by a infinitesimal volume $d^{3}r$ is equal to the infinitesimal probability dP, that a system specified by Ψ, is located within that volume.

$dP = \Psi^*\Psi d^3r$

2.3   The probability that an energy system represented by the wave function Ψ, is somewhere in all space is unity, which is the basis for a normalized wave function.

$P = \int dP = \int \Psi^*\Psi d^3r = 1$

3. The Heisenberg Uncertainty Principle:

3.1   In an ideal experiment, the product of the standard deviation in the measurement of a system's momentum, Δp and the standard deviation in the measurement of its position, Δr must be greater than a non-zero constant. This constant is Planck's constant divided by four pi $(h/(4\pi) = \hbar/2)$ where $\hbar$ is Planck's constant divided by 2π.

$\Delta p \Delta r \geq \hbar/2$

This means that an energy system's position and momentum cannot be known simultaneously.

3.2   Another expression of the Heisenberg uncertainty principle is:

$$\Delta E \Delta t \geq \hbar/2$$

where ΔE is the standard deviation in the measurement of a system's energy and Δt is the standard deviation of the measured times that it had that energy. This means that a system's energy and when it had that energy cannot be known simultaneously.

4. **The energy of a photon (E_γ):** an electromagnetic wave's energy is either the product of its frequency ω, and Planck's constant $\hbar$, or its mass m_γ, and the speed of light, c squared.

$$E_\gamma = \hbar\omega = m_\gamma c^2$$

5. **De Broglie's relationship:** which expresses that the wavelength of a particle λ is inversely proportional to its momentum. The constant of proportionality is Planck's constant, h.

$\lambda = h/mv$

and is sometimes written as

$\lambda\mkern-11mu{}^{-} = \hbar/mv$

where $\lambda\mkern-11mu{}^{-} = \lambda/(2\pi)$ and $\hbar = h/(2\pi)$

6. **Einstein's Laws of Special Relativity:** The first four relativistic laws are derived by assuming that the velocity of light c, is independent of the velocity of the source of light as well as the velocity of the observer.

6.1 A system's mass m increases if it is moving with a velocity v compared to the velocity of light c, in vacuum. Initially when the system had a velocity of zero, its rest mass is m_0.

$$m = m_0 \left(1 - (v/c)^2\right)^{-1/2}$$

6.2 A system's length ℓ decreases if it is moving with a velocity v compared to the velocity of light c, in vacuum. Initially when the system had a velocity of zero, its rest length is ℓ_0. ℓ is in the direction of the velocity.

$$\ell = \ell_0 \left(1 - (v/c)^2\right)^{1/2}$$

6.3   A system's clock time length, t slows (stretches) if it is moving with a velocity v compared to the velocity of light c, in vacuum. Initially when the system was at rest (had a velocity of zero), it had a clock time length of t_0.

$$t = t_0 \left(1 - (v/c)^2\right)^{-1/2}$$

6.4   The total mechanical energy E of a system containing mass is the product of its mass m and the square of the velocity of light c.

$$E = mc^2$$

where $m = m_0(1 - (v/c)^2)^{-1/2}$ is dependent on its velocity v. m_0 (rest mass) is its mass when $v = 0$.

6.5 The relativistic kinetic energy T, of a system in motion is the difference (between its mass in motion less its rest mass) times the velocity of light c, squared.

Einsteinian: $T_E = (m - m_0)c^2$

where $m = m_0(1 - (v/c)^2)^{-1/2}$ is dependent on its velocity, v. For small velocities compared to the velocity of light, the Einsteinian kinetic energy reduces to the Newtonian kinetic energy as a first order approximation. For $v \ll c$, $(m - m_0)c^2 \cong (\frac{1}{2})m_0v^2$

Newtonian: $T_N = (\tfrac{1}{2})m_0v^2$

7. Laws of Thermodynamics

7.1   The first law of thermodynamics says that within a closed (isolated) system an amount of heat added to the system dQ results in an increase in its internal energy dU and an amount of work done, dW. Usually, dU results in an increase in internal temperature while dW results in a change in volume dV against a constant pressure p. This also means that energy is conserved for a closed system.

$$dQ = dU + dW \quad \text{where} \quad dW = pdV$$

7.2   The second law of thermodynamics says that a change in the entropy dS of a system undergoing a reversible process is defined to be the amount of heat added dQ divided by its temperature T. If the process is irreversible, then the entropy is always greater than the amount of heat added divided by its temperature.

$dS \geq dQ/T$

where the equality implies reversibility and the greater than symbol (>) implies irreversibility.

7.3 The perfect gas law says that the gas pressure p multiplied by the volume of gas V is proportional to the number of moles n of gas multiplied by the absolute temperature T of the gas. The constant of proportionality R is known as the universal gas constant.

$pV = nRT$

7.4 The fundamental law of heat conduction says that the rate of heat flow dQ/dt across a infinitely thin slab dx of material perpendicular to the surface of the slab is proportional to the surface area A of the slab and the instantaneous absolute temperature change per unit thickness dT/dx of the material. The constant of proportionality C_T is known as the thermal conductivity of the material. The minus sign means that heat flow is in a direction of decreasing temperature.

$$dQ/dt = -C_T A dT/dx$$

7.5   The internal energy U of an ideal gas containing N molecules is proportional to the product of N and the absolute temperature T. The constant of proportionality is $3k/2$ where k is Boltzmann's constant.

$$U = (3/2)NkT$$

7.6   In an idealized heated solid called a cavity radiator, the energy radiated from the cavity interior per unit area (called total cavity radiancy, R_C) is proportional to the fourth power of the absolute temperature T. The constant of proportionality σ is called the Stefan-Boltzmann constant.

$$R_C = \sigma T^4$$

8. **Temperatures and Conversions**

C^0 is the symbol for degrees Celsius, F^0 is the symbol for degrees Fahrenheit and K^0 means degrees Kelvin (Absolute).

8.1 Water freezes at 0 C^0 at standard atmospheric pressure.

8.2 Water boils at 100 C^0 at standard atmospheric pressure.

8.3 The triple point (existing simultaneously as a gas, liquid and solid) of water   occurs at a temperature of 273.16 K^0 and atmospheric pressure of 611.73 Pascals (Newtons per square meter).

8.4 $C^0/100 = (F^0 - 32)/180$

8.5 $K^0 = C^0 + 273.16$

Electromechanical Laws

1. **Maxwell's Equations**:

1.1   The source of the electric field ($\mathbf{E}$) is charge density ρ_q. $\nabla = (\partial/\partial x, \partial/\partial y, \partial/\partial z)$ is the normal vector operator, ($\bullet$) is the normal vector scalar product and ε_0 is a constant called the permittivity of free space. This law is also known as Gauss's law for electricity. The differential form is

$$\nabla \bullet \mathbf{E} = \rho_q/\varepsilon_0$$

The integral form is

$$\varepsilon_0 \oiint \mathbf{E} \bullet \mathbf{n} dS = q$$

where $\oiint$ means integration over the closed surface S, $\mathbf{n}$ is a unit vector normal to S enclosing the charge q.

1.1.1 Maxwell's first equation and may be used to derive Coulomb's law which states that the force between two charges is proportional to the product of the two charges and inversely proportional to the square of the distance between their charge centers. The force is in a direction on a line drawn between the two charges q_1 and q_2 denoted by the unit vector $\mathbf{r}_u$. $K_C = 1/(4\pi\varepsilon_0)$ will be called Coulomb's constant.

$$\mathbf{F} = \mathbf{r}_u K_C q_1 q_2 / r^2$$

giving rise to the electrical potential energy, V between q_1 and q_2

$$V = K_C q_1 q_2 / r$$

If the charges are both positive or both negative, the force is repulsive (like charges repel one another), otherwise the force is attractive (unlike charges attract one another).

1.2 The source of the magnetic field $\mathbf{B}$ is zero. This is Maxwell's second equation. This also means that magnetic fields always exist in closed loops and magnetic monopoles do not exist. This law is also known as Gauss's law for magnetism. The differential form is

$$\nabla \bullet \mathbf{B} = 0$$

The integral form is

$$\oiint \mathbf{B} \bullet \mathrm{ndS} = 0$$

where $\oiint$ means integration over any closed surface S, $\mathbf{n}$ is a unit vector perpendicular to the surface, S.

1.3 Ampere's law is also known as Maxwell's third equation. Electrical current density $\mathbf{J}_i$ and/or dynamic electric fields, $\partial \mathbf{E}/\partial t$ give rise to circulating magnetic fields ($\mathbf{B}$). μ_0 is known as the permeability constant of free space. The differential form is

$$\nabla \times \mathbf{B} = \mu_0 \mathbf{J}_i + \mu_0 \varepsilon_0 \partial \mathbf{E}/\partial t$$

where $\nabla = (\partial/\partial x, \partial/\partial y, \partial/\partial z)$ is the Del vector operator and $\mathbf{x}$ is the vector cross product. The integral form is

$$(1/\mu_0) \oint \mathbf{B} \bullet \mathbf{ds} = i$$

where $\oint$ means integration over a closed line s, circulating around the electrical current, i. $\mathbf{ds}$ is an infinitesimal vector line element of s, that $\mathbf{B}$ circulates through. $\mathbf{B}$ is perpendicular to the direction of the electrical current i.

1.4   Faraday's law is also known as Maxwell's fourth equation. Dynamic magnetic fields, $(\partial \mathbf{B}/\partial t)$ give rise to circulating electric fields ($\mathbf{E}$). The differential form is

$$\nabla \times \mathbf{E} = -\partial \mathbf{B}/\partial t$$

where $\nabla = (\partial/\partial x, \partial/\partial y, \partial/\partial z)$ is the normal Del vector operator and $\mathbf{x}$ is the vector cross product.

The integral form is

$$\oint \mathbf{E} \bullet \mathbf{ds} = -\iint (\partial \mathbf{B}/\partial t) \bullet \mathbf{n}dS = -\partial \Phi/\partial t$$

where $\Phi = \iint \mathbf{B} \bullet \mathbf{n}dS$ is called the magnetic flux in which $\mathbf{B}$ penetrates the surface area S. $\mathbf{n}$ is a unit vector perpendicular to the surface area S.

2. The Lorentz Force:

The force $\mathbf{F}$ on a charge q moving with velocity $\mathbf{v}$ by an external electric field $\mathbf{E}$ and by an external magnetic field $\mathbf{B}$ and $\mathbf{x}$ is the normal vector cross product.

$$\mathbf{F} = q\mathbf{E} + q\mathbf{v} \ \mathbf{x} \ \mathbf{B}$$

3. Electromagnetic Wave Equations:

When there is no charges or currents, as in the vacuum of matter free space, Maxwell's equations yield a wave equation that is satisfied by both the electric field **E** as well as the magnetic field **B**. These equations yields the precise description of induced electromagnetic fields.

3.1 $\nabla^2\mathbf{E} - \partial^2\mathbf{E}/(c^2\partial t^2) = 0$ and

3.2 $\nabla^2\mathbf{B} - \partial^2\mathbf{B}/(c^2\partial t^2) = 0$

where $\nabla^2 = \nabla \bullet \nabla = \partial^2/\partial x^2 + \partial^2/\partial y^2 + \partial^2/\partial z^2$, t is the time and c is the speed of light in vacuum. Note that if one utilizes the gradient operator, $\square$ defined as $\square = (\partial/\partial x, \partial/\partial y, \partial/\partial z, \partial/\partial(ict))$ then, the Dalembertian operator, $\square^2 = \square \bullet \square = \partial^2/\partial x^2 + \partial^2/\partial y^2 + \partial^2/\partial z^2 - \partial^2/(c^2\partial t^2)$ simplifies the electromagnetic wave equations 3.1 and 3.2 as

3.1.1 $\square^2\mathbf{E} = 0$ and

3.2.1 $\square^2\mathbf{B} = 0$

Conservation Laws

1. Conservation of energy: A system's total energy, E_T is the same both before (B) and after (A) any energy transformation.

$(E_T)_B = (E_T)_A$

2. Conservation of momentum: A system's total momentum, p_T is the same both before and after any energy transformation.

$(p_T)_B = (p_T)_A$

3. Conservation of angular momentum: A system's total angular momentum, L_T is the same both before and after any energy transformation.

$(L_T)_B = (L_T)_A$

4. Conservation of charge: A system's total charge, Q_T is the same both before and after any energy transformation.

$(Q_T)_B = (Q_T)_A$

5. Conservation of baryon number: A system's baryon number, N_B is the same both before and after any energy transformation. Baryons are composed of quarks. Quarks have baryon number +1/3. Anti-quarks have baryon number −1/3.

$(N_B)_B = (N_B)_A$

6. Conservation of lepton number: A system's lepton number, N_L is the same both before and after any energy transformation.

$(N_L)_B = (N_L)_A$

7. For any energy system, another related energy system predicted by the simultaneous operations of time reversal, charge conjugation (signs of all charges involved are reversed) and space reversal (mirror image or parity) is also possible. This is called CPT for short. Below, E_T is the total energy of a system and BCPT means before the CPT

operation and ACPT means after the CPT operation.

$$(E_T)_{BCPT} = (E_T)_{ACPT}$$

Basic Units

position: (measured with a ruler)

meter = m

mass: (measured with a balance scale)

kilogram = kg

time: (measured with a clock)

second = s

charge: (measured with a voltmeter)

coulomb = coul

Equivalent Units

Force: Newton = nt = $kg\text{–}m/s^2$

Pressure: Pascal = nt/m^2

Energy: joule = nt–m

Inductance: henry = $joule\text{–}m\text{–}s^2/coul^2$

Capacitance: farad = $coul^2/joule$

Basic Physical Constants

Name	Symbol	Value
Speed of light	c	3.00×10^8 m/s
Gravitational Constant	G	6.67×10^{-11} nt–m^2/kg^2
Avogadro's number	N_0	6.023×10^{23} /mole
Universal Gas Constant	R	8.32 joules/(mole–K^0)
(Planck's constant)/2π	$\hbar$	1.055×10^{-34} joule–s
Planck length	$L_P = (\hbar G/c^3)^{1/2}$	1.616×10^{-35} m
Planck time	$T_P = (\hbar G/c^5)^{1/2}$	5.391×10^{-44} s
Planck mass	$M_P = (\hbar c/G)^{1/2}$	2.177×10^{-8} kg
Boltzmann's constant	k	1.38×10^{-23} joules/(molecule–K^0)
Stefan-Boltzmann constant	σ	5.67×10^{-8} joules/m^2/(K^0)^4
Permeability constant	μ_0	1.26×10^{-6} henry/m

Name	Symbol	Value
Permittivity constant	ε_0	8.85×10^{-12} farad/m
Electron charge	q_e	-1.6022×10^{-19} coul
Electron rest mass	m_e	9.11×10^{-31} kg
Proton rest mass	m_p	1.67239×10^{-27} kg
Neutron rest mass	m_N	1.6747×10^{-27} kg
Coulombs constant	$1/(4\pi\varepsilon_0)$	8.99×10^{9} nt-m^2/coul^2

Basic Elementary Particles

Preliminary Particle Descriptors

1. Family Names – Particles belong to functional families having a set number of family members. For example, the gluon family has eight members and they function to provide the strong nuclear force that hold quarks together. Individual particles have both a historical name and a symbol. For example, an electron has the symbol e^-.

2. Color – Quarks can either be red, green or blue (r,g,b). Anti-quarks can either be –red, –green or –blue (–r, –g, –b). This is similar to charge coming in two types, the minus (–) and the plus (+) type.

3. Charge – measured in units of positive electronic charge or the charge on a positron (anti–electron). The charge magnitude of a negative electron (e^-) or a positive positron (e^+) are equal. An anti-particle has the opposite charge as the particle.

4. Spin – Axial angular momentum measured in units of Planck's constant divided by 2π and denoted by $\hbar$. Quantum Spin is specified as positive, but it is understood that quantum mechanically, it can either be positive (parallel) or negative (anti–parallel) to any given direction. Fermions (matter particles) have half integral values of $\hbar$. Bosons (force field particles) have integral values of $\hbar$.

5. Helicity – Helicity is also given in terms of $\hbar$ and may be thought of as the component of the particle's spin in the direction of the particle's velocity vector. The helicity of particles moving at the velocity of light is different than the helicity of particles that do not. Particles moving at the velocity of light, c such as photons, must have zero rest mass and there is no coordinate system for which its velocity is zero. Thus, the component of a photon's spin ($\hbar$) along its velocity vector is the same as its spin orientation, either $+\hbar$ or $-\hbar$ since it cannot be observed at rest. Thus, a photon has an intrinsic helicity the same as

its intrinsic spin. On the other hand, particles with non-zero rest mass have non–intrinsic helicity dependent on the observer since their spin can be observed when they are at rest and their spin components in the direction of motion must have a quantum difference of $+\hbar$. For example, the weakons, responsible for the electroweak forces, with non-zero rest masses and spin of $\hbar$ have helicity of either $-\hbar$, 0, or $+\hbar$. A particle and its anti-particle have opposite helicity.

6. Rest Mass – Measured in either Proton rest masses (Mp) or millions of electron volts (Mev). An electron volt (1.602×10^{-19} joules) is the kinetic energy an electron gains by being propelled a distance of one meter by an electrical field of strength, one volt per meter. The equivalent energy of a proton at rest is 938 Mev. The reason rest mass can be measured in terms of energy is because of Einstein's famous equation $E_0 = m_0 c^2$ which relates rest mass, m_0 to rest mass energy, E_0 by a constant, being the square of the speed of light, c^2.

7. Field Energy – Force fields are caused by corresponding field particles having integral values of $\hbar$ (called bosons). Matter particles having half integral values of $\hbar$ (called fermions) are influenced by force fields caused by their interaction with the corresponding boson. The four force fields are strong nuclear (gluons), electroweak (weakons), electromagnetic (photons) and gravitational (gravitons).

Anti-Particle Properties

All particles have an anti-particle. The anti-particle has the opposite charge of the particle. The anti-particle has the opposite helicity of the particle. The anti-particle of a non-zero rest mass particle having zero charge, and having a spin of one $\hbar$ and zero helicity is the particle itself. The anti-particle has the same mass as the particle. A particle and its anti-particle (that is not itself) annihilate one another upon contact in a burst of other energetic particles.

Matter Energy Particles

All material energy is composed of fundamental matter particles experimentally observed to exist as three energy families (UP, CHARMED, TOP) of four fermions each, in its simplest representation. Two of the fermions are light and are called leptons and two of the fermions are heavy and are called quarks. One of the leptons carries a negative electronic charge, the other has no charge.

Origin of the UP Family

The nuclei of atoms are composed of neutrons and protons. A neutron consists of two (red and blue) down quarks, ($d_R^{-1/3}$, $d_B^{-1/3}$, $u_G^{2/3}$) and one (green) up quark, . A proton consists of two (red and blue) up quarks, and one (green) down quark, ($u_R^{2/3}$, $u_B^{2/3}$, $d_G^{-1/3}$). Any other cyclic permutation of red, green or blue colored quarks in neutrons or protons is possible. The proton is stable. An isolated neutron, n is unstable and will decay into a proton, p electron, e^- and an electron anti-neutrino, $\acute{v}_e$. The net effect is that one of the down quarks of the neutron will change into an electron, anti-neutrino and an up quark. This effectively transformed the internal structure of a neutron ($d_R^{-1/3}$, $d_B^{-1/3}$, $u_G^{2/3}$) into that of a proton ($u_R^{2/3}$, $u_B^{2/3}$, $d_G^{-1/3}$). The up quark has a charge of 2/3 e^+ while the down quark has a charge of $-1/3$ e^+. Thus a proton has a net charge of e^+ while the neutron has a net charge of 0. The UP family making up neutrons and protons consist of four family members which are the up quark, down quark, electron and its anti-neutrino. There are two other four member families. The TOP family has the highest rest mass energy particle members. The CHARMED family has intermediate rest mass energy particle members. The UP family has the lowest rest mass energy particles. Each family maintains the same relationships between its members.

The UP Family

The UP family consists of an up quark, $u^{2/3}$, a down quark, $d^{-1/3}$, electron, e^-, with its electron anti-neutrino, $\acute{\upsilon}_e$. The quarks can either be red, blue or green. The up quark has a charge of 2/3 e^+ while the down quark has a charge of $-1/3$ e^+. The electron has a rest mass energy of .511 Mev. These particles have the lowest rest mass energy and represent the ground state rest mass energy of the matter families. All UP fermion family members have a spin of $\frac{1}{2}\hbar$ and a helicity of plus or minus $\frac{1}{2}\hbar$.

The CHARMED Family

The CHARMED family consists of a charmed quark, $c^{2/3}$, a strange quark, $s^{-1/3}$, muon, μ^- with its muon anti-neutrino, $\acute{\upsilon}_\mu$. The quarks can either be red, blue or green. The charmed quark has a charge of 2/3 e^+ while the strange quark has a charge of $-1/3$ e^+. The muon has a rest mass energy of 105.66 Mev. These particles have intermediate energy and represent a higher rest mass energy state than the UP family. All CHARMED fermion family members have a spin of $\frac{1}{2}\hbar$ and a helicity of plus or minus $\frac{1}{2}\hbar$.

The TOP Family

The TOP family consists of a top quark, $t^{2/3}$, bottom quark, $b^{-1/3}$, tauon, τ^- with its tauon anti-neutrino, $\acute{\upsilon}_\tau$. The quarks can either be red, blue or green. The top quark has a charge of 2/3 e^+ while the bottom quark has a charge of $-1/3$ e^+. The tauon has a rest mass of 1784.2 Mev. These particles have the highest rest mass energy state and represent a higher energy state than that of the CHARMED family. All TOP fermion family members have a spin of $\frac{1}{2}\hbar$ and a helicity of plus or minus $\frac{1}{2}\hbar$.

Field Energy Particles

Gluon Family

Gluons ($g_1 - g_8$) are responsible for the strong force field between the three colored (red, green and blue) quarks making up protons and neutrons, of which all nuclei are composed. There are eight different gluons. Gluons carry color combinations (r, g, b, −r, −g, −b) and compose the gluon field holding quark trios together in protons and neutrons. Gluons have a spin of $\hbar$. Gluons have zero rest mass and therefore move at c, the velocity of light. Thus, gluons have helicity of either plus or minus $\hbar$.

Photon Family

Photons (γ) are responsible for the electromagnetic forces which act between charges. Photons have no color and no charge. Photons have a spin of $\hbar$. Photons have no rest mass and move at the velocity of light. Thus, photons have helicity of either plus or minus one $\hbar$. The positive helicity photon is the anti-photon of the negative helicity photon. While in flight, photons have mass, energy and momentum.

The Weakon Family

Weakons give rise to the electroweak force field responsible for radioactive decay. Recall that a neutron is composed of two down quarks and one up quark. The decay of an isolated neutron is an example of radioactive beta (electron) decay in which one of the down quarks in a neutron decays into a weakon (the omega minus) which then decays into an up quark, electron and anti-neutrino. The net effect is that a neutron decays into a proton, electron and anti-neutrino. There are three different weakons, the omega minus (Ω^-), omega zero or zeta (Z^0) and the omega plus (Ω^+). These weakons have no color and carry charges of e^-, 0, e^+ respectively. Weakons have a spin of $\hbar$ and each can have helicity of $-\hbar$, 0 or $+\hbar$. Weakons have rest masses of 85 Mp, 260 Mp and 85 Mp respectively. Anti-weakons have opposite charges and helicities as the corresponding weakons.

The Meson Families

Mesons give rise to the forces between baryons (quark trios). Mesons are not elemental but are composed of quark anti-quark pairs (combos taken from any of the three families of quarks) and are mentioned here for completeness. Obviously, there are many families of mesons, and the pi meson family (pions) are responsible for forces between nucleons (either neutrons or protons). Pions will be presented next as an example.

The Pi Meson (Pion) Family

The Pions are responsible for forces between nucleons (either neutrons or protons) and are composed of quark anti-quark pairs. The pi minus (π^-) is composed of a down quark with a charge of $-1/3$ e^+ and an anti-up quark with a charge of $-2/3$ e^+ for a total charge of e^-. The pi zero (π^0) is a mixture of an up quark and an anti-up quark, with a down quark and an anti-down quark. The pi plus (π^+) is composed of an up quark with a charge of $+2/3$ e^+ and an anti-down quark with a charge of $+1/3$ e^+ for a total charge of e^+. These pions have no color and carry charges of e^-, 0, e^+ respectively. Pions have a spin of 0 and each has helicity of 0. The charged pions have rest masses of 139.57 Mev, while the pi zero has a rest mass of 134.96 Mev. The anti-pi minus is the pi plus. The anti-pi plus is the pi minus. The anti-pi zero is the pi zero itself.

Graviton Family

Gravitons (G_- and G_+) are responsible for the gravitational force fields which act between masses. Gravitons have no color and no charge.

The G_- graviton has a spin of $2\hbar$ and a zero rest mass. It moves at the velocity of light and thus, its helicity is $-2\hbar$ or $+2\hbar$. It is assumed to have negative mass in flight while being exchanged between any two positive masses or any two negative anti-masses. This is because the gravitational potential energy between two positive masses or two negative masses is negative.

Because of a new scientific theory called "Nature of the First Cause", in which positive matter is gravitationally repelled by negative anti-matter, the G_+ graviton is postulated to exist. It also has a spin of $2\hbar$. It is assumed to have a zero rest mass and moves with the velocity of light and has positive mass in its flight between negative anti-matter and positive matter. Thus, the G_+ graviton also has helicity of $-2\hbar$, or $2\hbar$. The G_+ gravitons fill up all spacetime and are responsible for the force of repulsion between negative anti-matter and positive matter. By the "First Cause" theory, it makes up the repulsive gravitational field which is responsible for the accelerated expansion of distant positive matter in the universe (galaxies not in the local group).

The Higgs Family

There are two Higgs bosons (H_L and H_H) called the light Higgs boson, H_L of the unified electroweak theory and the heavy Higgs boson, H_H of the grand unified theory. The heavy Higgs boson, makes up the Higgs field and permeates all spacetime. This field is responsible for assigning masses to all the fundamental particles. The light Higgs boson is responsible for assigning the masses to the weakons. Both Higgs bosons have a spin of zero (0), and thus they both have a helicity of zero. Both Higgs bosons have non–zero rest masses with the light Higgs rest mass at roughly 10^5 Mev and the heavy Higgs rest mass of about 10^{17} Mev.

Complete Set of Particles

All matter particles which have been discovered are combinations of the above elementary matter particles. All the known force fields consists of varying energy and intensity of the above force field particles.

The Hadrons (consisting of quarks) which are matter particles that have been discovered now number over two hundred exceeding the number of known elements.

References

Al-Khalili, Jim, *Quantum, A Guide for the Perplexed*, United Kingdom, Weidenfeld & Nicolson, 2003

Ames, Joseph Sweetman & Murnaghan, Francis D., *Theoretical Mechanics An Introduction to Mathematical Physics*, New York, Dover Publications, Inc., 1957

Atkins, K. R., *Physics*, New York, John Wiley & Sons, Inc., 1965

Bennett, Jeffrey & Donahue, Megan & Schneider, Nicholas & Voit, Mark, *The Cosmic Perspective*, New York, Addison Wesley, 2004

Bergmann, Peter Gabriel, *Introduction to the Theory Of Relativity*, New York, Dover Publications, Inc., 1976

Blass, Gerhard A., *Theoretical Physics*, New York, Appleton-Century-Crofts, 1962

Bova, Ben, *The Fourth State of Matter*, New York, New American Library, Inc., 1974

Born, Max, *Einstein's Theory of Relativity*, New York, Dover Publications, Inc., 1962

Breithaupt, Jim, *Cosmology*, Blacklick, OH, McGraw-Hill, 1999

Davies, Paul, *The New Physics*, New York, Cambridge University Press, 1996

De Broglie, Louis, *matter and light*, New York, Dover Publications, Inc., 1939

Einstein, Albert, *Builders of the Universe*, Los Angeles, CA, U. S. Library Association, Inc., 1932

Einstein, Albert, & Lorentz, H. & A., Minkowski, H., & Weyl, H., *The Principle of Relativity*, New York, Dover Publications, Inc., 1952

Einstein, Albert, *Relativity The Special and General Theory*, New York, Crown Publishers, Inc., 1961

Fermi, Enrico, *thermodynamics*, New York, Dover Publications Inc., 1956

Feynman, Richard P., *QED The Strange Theory of Light and Matter*, Princeton, New Jersey, Princeton University Press, 1988

Feynman, Richard P., *Six Not So Easy Pieces*, New York, Basic Books, 1997

Frankel, Theodore, *Gravitational Curvature An Introduction to Einstein's Theory*, San Francisco, W. H. Freeman and Company, 1979

Gamow, George, *Gravity*, New York, Dover Publications, Inc., 2002

Goldstein, Herbert, *Classical Mechanics*, London, Addison-Wesley Publishing Company, Inc., 1950

Greene, Brian, *The Elegant Universe: Superstrings, Hidden Dimensions, and the Quest for the Ultimate Theory*, New York, W. W. Norton, 1999

Guth, Alan H., *The Inflationary Universe: The Quest for a New Theory of Cosmic Origin*, Perseus Books, 1997

Halliday, David & Resnick, Robert, *Physics For Students of Science and Engineering*, New York, John Wiley & Sons, Inc., 1962

Hawking, Stephen & Penrose, Roger, *The Nature of Space and Time*, New Jersey, Princeton University Press, 1996

Heisenberg, Werner Karl, *The Nature of Elementary Particles*, in Physics Today, Page 39, March 1976

Kaku, Michio, *Hyperspace*, New York, Anchor Books, 1995

Kaku, Michio, *Parallel Worlds*, New York, Anchor Books, 2006

Kaplan, Irving, *Nuclear Physics*, Reading, Massachusetts, Addison-Wesley Publishing Company, Inc., 1962

McMahon, David, *quantum mechanics demystified*, New York, McGraw Hill, 2005

Messiah, Albert, *Quantum Mechanics*, New York, Dover Publications, Inc., 1999

Musser, George, *Growing Pains*, Scientific American, Page 32, July 2004

Ostriker, Jeremiah P., & Steinhardt, Paul J., *The Quintessential Universe*, Scientific American, Page 46, January 2001

Park, David, *Introduction to the Quantum Theory*, New York, McGraw-Hill Book Company, 1964

Peebles, P. J. E., *Principles of Physical Cosmology*, Princeton, New Jersey, Princeton University Press, 1993

Penrose, Roger, *The Road To Reality, A Complete Guide to the Laws of the Universe*, New York, Vintage Books, 2004

Powell, John L. & Crasemann, Bernd, *Quantum Mechanics*, Reading, Massachusetts, Addison-Wesley Publishing Company, Inc., 1961

Ridpath, Ian, *The Illustrated Encyclopedia of the Universe*, New York, Watson-Guptil Publications, 2001

Riggs, Shelton, *An Alternative Lorentz Invariant, Relativistic Wave Equation,* Version 7.7, El Paso, Texas, www.co.el-paso.tx.us/clerk/deedsearch.htm, instrument number 20060091478, El Paso County Courthouse, 2006

Sears, Francis W., & Zemansky, Mark W. & Young, Hugh D., *College Physics*, Menlo Park, California, Addison-Wesley Publishing Company, 1986

Segre, Emilio, *Nuclei and Particles*, New York, W. A. Benjamin, Inc., 1964

Shortley, George & Williams, Dudley, *Elements of Physics For Students of Science and Engineering*, Englewood Cliffs, New Jersey, Prentice-Hall, Inc., 1965

Van Heuvelen, Alan, *Physics, A General Introduction*, Boston, Little, Brown and Company, 1982

Weinberg, Steve, *Dreams of a Final Theory: The Search for the Fundamental Laws of Nature*, New York, Pantheon Books, 1992

Weld, LeRoy D., *A Textbook of Heat*, New York, The Macmillan Company, 1948

Weyl, Hermann, *Symmetry*, Princeton University Press, 1952

Young, Hugh D., *Statistical Treatment of Experimental Data*, McGraw-Hill Co., Inc., 1962

INDEX

A

C

D

E

F

H

I

N

S

U

V

W

Z

www.ingramcontent.com/pod-product-compliance
Lightning Source LLC
Chambersburg PA
CBHW071406170526
45165CB00001B/195